Tell Me Where It Hurts

Tell Me Where It Hurts

THE NEW SCIENCE OF PAIN AND HOW TO HEAL

Rachel Zoffness, PhD

GRAND CENTRAL

New York   Boston

This book is not intended as a substitute for medical advice of physicians or other healthcare providers. The reader should regularly consult a physician and their healthcare team in all matters relating to his or her health, and particularly in respect of any symptoms that may require diagnosis or medical attention.

Copyright © 2026 by Rachel Zoffness, PhD

Cover design by Peter Garceau.
Cover copyright © 2026 by Hachette Book Group, Inc.

Hachette Book Group supports the right to free expression and the value of copyright. The purpose of copyright is to encourage writers and artists to produce the creative works that enrich our culture.

The scanning, uploading, and distribution of this book without permission is a theft of the author's intellectual property. If you would like permission to use material from the book (other than for review purposes), please contact permissions@hbgusa.com. Thank you for your support of the author's rights.

Grand Central Publishing
Hachette Book Group
1290 Avenue of the Americas, New York, NY 10104
grandcentralpublishing.com
twitter.com/grandcentralpub

First Edition: March 2026

Grand Central Publishing is a division of Hachette Book Group, Inc. The Grand Central Publishing name and logo is a trademark of Hachette Book Group, Inc.

The publisher is not responsible for websites (or their content) that are not owned by the publisher.

The Hachette Speakers Bureau provides a wide range of authors for speaking events. To find out more, go to hachettespeakersbureau.com or email HachetteSpeakers@hbgusa.com.

Print book interior design by Jeff Stiefel

Library of Congress Cataloging-in-Publication Data has been applied for.

ISBNs: 978-1-5387-5814-4 (hardcover), 978-1-5387-5815-1 (ebook)

Printed in Canada

MRQ-T

10 9 8 7 6 5 4 3 2 1

To Ferdinand, my muse.

*To every person living with pain,
and the brave army of healthcare providers who treat them:
You are not alone. This book is dedicated to you.*

Ask not what disease the person has,
but rather what person the disease has.
—*Dr. William Osler*

There are two ways of spreading light:
to be the candle, or the mirror that reflects it.
—*Edith Wharton*

Contents

Foreword 1
Introduction and Welcome 3

Part I: Pain Isn't What You Think It Is

1: Everything You've Been Told About Pain Is Wrong 9
2: The Real Science of Pain:
How Pain Works and Why We Have It 22
3: If the Brain Can Change, Pain Can Change 52
4: The Pain Recipe 68

Part II: The Pillars of Pain

5: Pain Is Emotional: Reconnecting Brain and Body 79
6: Pain Is Cognitive: The Mind, the Brain, and Pain 96
7: Pain Is Social: Why Friends Are *Actually* Medicine 116
8: Pain Is Environmental: The Seed Is Only as Good as Its Soil 131
9: The Body's Pharmacy 149

Part III: The Pain Protocol

10: Welcome to the Revolution: From Pain to Power 173
11: The Biology of Balance: Sleep, Diet, and Movement 182

12: The Emotional Health Protocol:
Why EQ Matters in Medicine 211

13: The Power of Changing Your Mind:
Cognitive Strategies for Pain 239

14: Social Medicine: Bloom Where You're Planted 268

Conclusion: The Future of Pain Medicine Is You *291*
Acknowledgments *295*
Endnotes *297*

Foreword

I know pain.

As a neurosurgeon for the past fifty years, I've served as faculty at Yale University School of Medicine and New York Medical College, written and lectured extensively about pain, and treated thousands of patients. I myself have required surgery for my own pain. But my approach to medicine has had a very unique trajectory.

As a young medical student at Beth Abraham Hospital, my mentor was a junior doctor unknown to me—and, at that time, unknown to the public—named Dr. Oliver Sacks. He became my teacher and friend and, ultimately, an acclaimed author and world-renowned speaker. My training with him turned out to be one of the most remarkable experiences of my medical education. His books continue to exemplify his singular approach to medicine.

With great patience and an abundance of joy, Dr. Sacks explored the inner lives, struggles, and triumphs of his patients. Each patient had a unique story. That story was the window that revealed the missing clues—and to open that window was to treat the patient as a whole person. Reading Dr. Zoffness's book, I thought again of Dr. Sacks and his work. This book similarly transforms stories about pain into stories about *people*.

While most books about pain address the pain of a specific body part or offer "easy cures," *Tell Me Where It Hurts* stands out as unique. Here, for the first time, is a book that brilliantly integrates pain neuroscience with the individual. It is evidence-based, thoroughly researched, readily accessible, and easily understood.

Dr. Zoffness is a highly skilled, respected clinician who has mastery of the vast, complicated topic of pain. She is also the consummate educator. On every page, I felt that she was speaking to me. Like the patients in her book, I felt that she understood *my* pain story, my unique history. She allowed me to finally understand the science of my own pain and the many factors that influence it.

I urge you to read this important book. In it, you will discover your own pain story—and, perhaps for the first time, experience the window opening, and find those missing clues revealed.

—Jack Stern, MD, PhD, associate professor of neurosurgery at New York Medical College, former faculty at Yale University School of Medicine and Weill Cornell School of Medicine, author of *Ending Back Pain*

Introduction and Welcome

I started studying pain for one main reason: I was scared of it.

At Brown University, I majored in Human Biology Brain and Behavior—a combination of neuroscience, psychology, and biology—and wrote my honors thesis on pain science under the tutelage of a renowned pain neuroscientist. My hope was that by understanding pain, my fear would fade. Over time, it actually did. But something unexpected also happened.

The more I learned, the more alarmed I became by how wrong we've been about pain—how little we're told about how it really works, how abysmally we treat it, and how dire the consequences. Because the truth about pain is that it is eminently understandable—*and also treatable.*

My name is Rachel Zoffness. I'm a pain scientist, pain psychologist, and faculty at the UCSF School of Medicine. I also teach the next generation of medical doctors at Stanford University. I trained at Brown, Columbia, the University of California San Diego, Rockefeller University, and Mount Sinai Hospital. I've served on the boards of America's leading pain organizations, consult on the development of pain management programs around the world, and give trainings for physicians at hospitals across the US.

When I finally opened my private practice dedicated to treating people living with pain, most of my patients had been deemed "untreatable medical mysteries." They'd tried every medication and intervention under the sun. But as we worked together, I watched as the seemingly impossible happened: One by one, they got out of bed and went back to life. You will read some of their anonymized stories in this book. *How could this possibly be?* I wondered. *I am not a magician.* What was different about my approach that helped my patients heal?

You will find those answers here—and more.

My deepest hope is that you glean something from these pages that helps you on your journey—whether you're someone living with pain, someone who treats it, a lover of science, or simply love someone who's suffering. Science offers incredible wisdom and promise, and there is *so much hope* to be found here. Join me as we completely revolutionize everything you've been told, and sold, about pain. I look forward to seeing you on the other side.

Tell Me Where It Hurts

PART I

Pain Isn't What You Think It Is

1

Everything You've Been Told About Pain Is Wrong

Migraines, Two Proms, and a Haircut

When I first met Sam, he'd been bedridden for four years. His skin was pasty and pale, his hair long and unwashed. He walked hunched over, his arms locked around his abdomen. When he sat on my couch, he rocked himself back and forth, his face contorted with pain. He'd been diagnosed with amplified body pain of unknown etiology and chronic migraine that caused daily, crippling headaches. Sam was so incapacitated that he'd had to drop work, study, exercise, and socializing, his life stripped down to the studs. He'd seen fourteen doctors and been on forty medications. Nothing had worked. In truth, that didn't particularly surprise me; I saw cases like this all the time. What surprised me was his age.

Sam was only sixteen years old.

Unable to attend school since seventh grade, Sam had fallen behind socially and academically, and was now completely disconnected from his community. He had no friends, no life, and no hope. After I gave a talk on pain science at a San Francisco hospital, Sam's mother approached me. Her eyes were frantic. "Please," she begged, desperate. "You have to help my son." They'd tried everything, she said; all other treatments had failed, and they were out of options. I wanted to help, but wondered if I could: If a top-tier hospital hadn't cured him, could I...? But I was the last stop on the train. How could I refuse?

I've been studying pain neuroscience, human biology, and clinical psychology for decades. But despite my science background, I actually think of myself as a pain detective. By the time a patient finds me, they're typically in Sam's position—exhausted, demoralized, and desperate for the pain to end. They've seen dozens of doctors and specialists, taken cabinets-ful of medication, and have been assigned an assortment of diagnoses. It's my job to examine all the clues and figure out what's really going on—the full constellation of factors triggering and perpetuating the pain, and what I need to do to treat it. I told Sam that I could help him if he was willing to trust me and try the medicine I was offering, even though—for the first time in his life—it wasn't a pill.

The first step was to establish rapport and dig for clues. A shy, private teenager, Sam was understandably hesitant to let me into his world. He gave one-word answers, shifting uncomfortably and looking at the floor. So instead of asking personal questions, I inquired about his hobbies. He loved two things, he said: soccer and reading. I told him that I'd been a shy child who loved books, too. The following week, Sam surprised me by bringing his favorite book for me to read, *Ready Player One*. I consumed it in a matter of days. We spent our next visit talking about the book, why he loved it, and his favorite characters—all of whom were lonely and socially isolated. It seemed telling. Slowly, we developed a relationship. He eventually disclosed that he'd been struggling with crippling social anxiety, depression, and suicidality for years—but that this had somehow been discarded by his team as unrelated to his pain, and thus left untreated. His body hurt; what could emotions possibly have to do with it?

Once my investigation was complete and I had a fuller picture of his life, family, stressors, triggers, and goals, we began treatment. The first step was teaching him the science of how pain works and how to heal. This required doing the careful dance I always do: helping him connect physical symptoms with emotional health without inadvertently conveying the message that the pain was "all in his head." It wasn't; it never is.

But this connection carries the built-in suggestion that a patient is "faking" or "mentally ill," a stigma that's as poisonous as it is pervasive. It is therefore something I always explicitly address.

After we discussed my approach to treatment, we began the action phase: helping Sam get out of bed and back to life, one tiny step at a time. Each week, I taught Sam some techniques to reduce his pain and manage his symptoms, and sent him home with a small task.

Week 1, he stood on his porch in the sun. Week 2, he walked to the corner mailbox. Week 3, he walked around the block and texted one friend. Week 4, he took his dog to the dog park and talked to a few strangers. Pain flared, then receded. Sam, his parents, and I discussed Sam's sleep and diet, and crafted a plan to help him alter his lifestyle. He stopped staying up until 4 a.m. playing videogames, set his alarm in the morning, and introduced fruits and vegetables into his largely bread-and-pasta diet. He tracked his progress on his phone, and each week had something promising to report: a new friend at the dog park, a reconnection with his former soccer coach, a pain-free walk.

Slowly, slowly, the physical, cognitive, and social activity began to change his brain. The more Sam did, the more he realized he *could* do. He began jogging—just a few minutes at first, stopping to rest when pain flared—and then a mile a day. As his anxiety, isolation, and depression abated, so did his pain. After two months, Sam was strong enough to get a tutor. He cracked open his math textbook and started catching up in school. He walked to the barber and got his first haircut in four years. He got glasses to ease eyestrain, and bought some new clothes. You should have seen the pride on Sam's face the day he showed up at my office wearing new sneakers, a new backpack, and an enormous grin. He was a teenager transformed. Sam spent the entire session excitedly showing off his books, pencil case, and school schedule.

When Sam finally returned to high school—fitter, stronger, healthier—he was asked to prom by not one girl, but two. He said yes to both. On the day of his graduation, Sam got onstage—this teen formerly paralyzed by anxiety, hopelessness, and pain—and announced that if you'd told him four years ago that he'd be graduating from high school, he'd never have believed you.

Sam's neurologist, an established physician who serves on the board of multiple prestigious medical journals, was baffled by his remarkable progress, calling it "a miracle." "What magic purple pill did you give that kid?" he asked incredulously. When I explained that treatment hadn't been a pill, but had instead targeted Sam's emotional and social pain in addition to the physical, he accused me of "stigmatizing" Sam. He argued that anxiety and depression had nothing to do with migraine, and stopped referring patients to me—even though he'd been treating Sam with pills and procedures for four years without any improvement.

I. The Painful Truth

Pain by the Numbers

Sam's story, while extreme, is a classic example of how we continue to mistreat pain—and the reason we continue to suffer.

Pain is part of the human condition. Not a single one of us will escape. Pain is coming for all of us, if it hasn't already—in the form of an injury, the pain of childbirth, back pain, or simply inhabiting an aging body. Many of us are managing pain right now; after all, pain is a symptom of nearly every known disease and illness. In fact, approximately 1.9 billion people around the globe currently live with chronic pain, 100 million in the US alone.[1] Should these numbers seem high, think again—by most accounts, these are likely *under*estimates. Pain is ubiquitous, notoriously difficult to measure, and commonly treated as a symptom of other underlying conditions like arthritis, cancer, or migraine. Moreover, these numbers don't include those affected by acute, short-term pain; people living in long-term care facilities; adults in the military; or incarcerated individuals. An immense financial burden, pain costs the US over $635 billion annually in medical costs and lost work productivity. And while you might assume chronic pain is reserved for the elderly, it impacts children, too: pediatric chronic pain, which includes common conditions like headaches and stomachaches, affects one in three children. An equal opportunist, pain impacts all genders, all races, and all cultures around the world. No one is exempt.

Our Treatments Don't Work

Given how common pain is, it's remarkable how terrible we are at treating it. Over the past several decades, a glaring disparity has emerged between pain prevalence and pain management: simply put, our treatments don't work. Rather than decreasing, chronic pain is on the *rise*—affecting more people today than cancer, heart disease, and diabetes combined.[2] Pain remains the number one reason for doctor's visits. Meanwhile, thousands die annually of overdoses on opioids, the pain pill of choice, which is still prescribed for everything from a toothache to low back pain. Yet despite the fact that the pharmaceutical industry is finally being taken to task to the tune of more than $50 billion in reparations and payouts for false advertising, this hasn't solved our pain problem.

It's not just medications that fall short. Attempts to cut, burn, ablate, inject, and excise pain are similarly rife with failure. Take back pain, for example. One of the most common types of chronic pain, it affects up to 85 percent of the American population. Back surgery is one of the most recommended and popular solutions. But this procedure is so frequently unsuccessful that failed attempts have earned their own formal medical diagnosis: failed back surgery syndrome. Medicine now openly acknowledges that back surgeries aren't the best solution for chronic back pain. In fact, they're no better than nonsurgical interventions—and are perversely associated with increased disability and long-term opioid use.[3]

Medical devices like spinal cord stimulators, implants that deliver electrical impulses to the spinal cord, fare no better, proving no more effective than placebo and carrying significant risk of harm.[4] Surgery-as-pain-cure for other painful conditions is equally as questionable: the *Journal of Pain Research* reports that chronic pain strikes up to 60 percent of patients *after* common operations.

Yet somehow, amid the disasters and the desperation, the treatment of pain remains a mystery—*despite the fact that it doesn't need to be.*

The Miseducation of Pain Medicine

One of the primary reasons pain continues to be misunderstood and mistreated is that the vast majority of us are never taught how pain works. Through no fault of our own, we aren't getting the pain education we need. Alarmingly, this includes the majority of healthcare providers across disciplines, including medicine, nursing, physical therapy, and occupational therapy. A recent survey reveals that 96 percent of medical schools in the US have zero dedicated, compulsory pain education.[5] Healthcare providers in other disciplines, like nursing and psychology, receive even less. That's not to mention patients—that is, the rest of us. We get none.

The select medical schools that do offer pain education commit fewer than five hours to the subject over four-plus years of medical training, relegating the subject to brief lectures and seminars.[6] Pain is typically a subtopic folded into courses such as anesthesia or pharmacology, rather than offered as a stand-alone, multidisciplinary course. When pain is taught, the focus is primarily, if not exclusively, on anatomy and physiology—the study of basic body parts and their functions.[7] Lest we mistakenly conclude that this reductive focus is restricted to medicine alone, it's also true of pain education across other healthcare disciplines, including physical therapy and chiropractic schools, which popularly assert that "the issue is in the tissues."

(Spoiler alert: It isn't.)

My own research confirms the same. Surveys of my students, medical residents and fellows at Stanford and University of California San Francisco—doctors who have already completed four years of medical school—reveal that they received little to no pain education over the course of their training.[8] The pain science class I teach, most report, is the first they've been offered.

To be clear, this isn't the fault of our healthcare providers. Across disciplines, providers have been clamoring for better pain education for years. Indeed, when primary care doctors are formally surveyed, 82 percent rate their medical school pain education as frustratingly insufficient. And two-thirds of our doctors—people who chose medicine so that they might help and heal—report that their education didn't adequately prepare them to treat chronic pain, leaving them feeling underprepared, burned out, and overwhelmed.

If pain science is rarely taught and is therefore rarely understood, it is suddenly much less mysterious why we don't seem to know how to treat it. As someone who either has pain now or will someday, this should concern you.

A lot.

The Blessing and Burden of Biomedicine

That adequate pain education is lacking across healthcare isn't news. For over a decade, this has been cited by experts as one of the primary barriers to effective pain treatment, and has even been referred to as a "crisis."[9] One of the reasons for this crisis is that our understanding of pain remains firmly rooted in the past.

The long-outdated biomedical model, upon which modern pain medicine rests, dates back to before 400 BCE. It is so old that it significantly predates Descartes, a philosopher who separated the mind from the body much like a cart uncoupled from its horse. This ancient conceptualization of human health is strictly concerned with anatomy and physiology, telescopically focusing on the body part that hurts rather than the whole person. Pain is thus blamed exclusively on injuries, posture, misalignment, "asymmetry," strains, and slipped discs.

Regardless of what ails you, you have likely been told the same.

Despite major advances in science and medicine revealing that pain is a complex neurobiological event impacted by thoughts, emotions, social health, and even our environment, pain continues to be incorrectly framed as a purely biomedical problem that requires a purely biomedical solution: that is, pills and procedures.[10] And when scans and tests reveal no physical abnormality, disease, or damage, patients are instructed to simply take painkillers and rest—or, worse, told the pain is "all in their heads." This overly simplistic paradigm stubbornly and dangerously ignores other well-established drivers of pain, including social, emotional, and environmental factors. By neglecting these, our prevailing model of health ultimately prevents appropriate care, hurting more than it helps.

The truth is that we know better. That pain is multifactorial—*not* just biological—is a scientific fact backed by many decades of research. I will spend the rest of this book proving it to you. Still, for a variety of reasons, many to do with our profit-driven healthcare system, pain medicine remains rooted in the antiquated biomedical model. We continue to be treated as disconnected body parts, despite being housed in one hyperconnected body.

The costs of this misguided approach make for grim reading. The United States now holds the dubious distinction of being one of the world's leading opioid prescribers. While we make up only 4.6 percent of the world's population, we have historically consumed over 80 percent of the world's opioid supply.[i] According to the US Centers for Disease Control, opioid-related deaths increased by 4,250 percent between 1979 and 2015. These deaths overlap with our pain crisis: More people in the US die from prescription opioid overdoses than from cocaine and heroin combined. And while our culture frequently stigmatizes heroin users as "addicts" and "mentally ill," 80 percent tragically report that their first opioid was a prescription pill.[11] What we've done to people living with pain—people like you and me—is no less than criminal.

This is not to disparage biomedicine, which has generated lifesaving solutions where before there were none. Biomedicine is the reason we have penicillin, vaccines, and drugs to fight cancer; the reason we have prosthetic limbs and heart transplants; and the reason we can now live until one hundred. But when it comes to *pain*, we miss the big picture when we focus solely on bones, bruises, and body parts. Because neuroscience

---

[i] These numbers vary depending upon who you ask. A more recent report from 2021 notes that "the United States has less than 5% of the world's population but consumed roughly 30% of the world's opioids in 2009, including more than 99% of the world's hydrocodone and 80% of the world's oxycodone." Sources: Duff, J. (2021). *Consumption of Prescription Opioids for Pain* (Congressional Research Service). Manchikanti, L. & Singh, A. (2008). Therapeutic opioids: A ten-year perspective on the complexities and complications of the escalating use, abuse, and nonmedical use of opioids. *Pain Physician*, *11*(2S), S63.

reveals that our brains matter, too—as do the environments we grow up in, the thoughts we think, and the foods we eat.

Driven by a broken healthcare system, the influence of Big Pharma, and poor pain education, pain medicine continues to fail patients and providers alike. Meanwhile, the discrepancy between the problem pain clearly presents, and the time we invest in teaching and treating it, continues to widen.

II. But We Can Fix It

Now for the good news. And there's a lot of it.

Thanks to recent advances in neuroscience and medicine, we now understand pain better than ever before. And with this knowledge comes a better understanding of how to heal. The true treatment of chronic pain isn't a magic pill, and it isn't a quick fix. But it also isn't a mystery.

We now know that pain is both more complex, and yet simpler and more intuitive, than we ever realized. At the crux of it are three major truths that fundamentally alter the landscape of pain medicine. The first is our concept of where pain is made. Tempting as it is to believe that pain is made exclusively by the body part that hurts, *that isn't actually true*. Rather, pain is constructed by the brain. This includes the parts of our brain that make emotions. In fact, research confirms that emotions are a critical part of the pain experience, constantly adjusting pain volume. This means pain—all pain—is always both physical *and* emotional. We all know this intuitively, having seen how our bodies ache more during times of stress and duress, how we can briefly forget about our pain when absorbed in a beloved activity, and how depression actually hurts.

The second important consequence of relocating pain is that neuroscience offers a walloping dollop of optimism. Studies reveal that the brain is "plastic," or ever-changing. The term for this is *neuroplasticity*, and it means that our brain cells have the ability to change the way they connect and communicate in response to new information. Each time you memorize a new route to work, for example, or rehearse a piece of music, the pathways in your brain rewire, creating new neural networks. In this way, brain structure and function constantly change throughout our lives with input, time, and experience. In fact, the brain continues to morph and change every day until the day we die. This is hopeful, encouraging news for people living with pain—because if the *brain* can change, *pain* can change. This is a mantra worth repeating.

The third crucial update to our understanding of pain requires finally abandoning the antiquated, outdated biomedical model. Pain, it turns out, is never just about bulging discs, broken bones, or bruised body parts. Rather, research shows that pain is biopsychosocial.[12,13,14,15,16,17] This means there are three equally important domains of human health that contribute to pain production (and reduction!): biological, psychological, and sociological. Counterintuitive as it may seem, thousands upon thousands of studies confirm that stress, trauma, mental health, social health, loneliness, lifestyle, and environment impact the pain we feel just as much as any broken bone. Our short-sighted focus on the "bio" alone means *we've been missing two-thirds of the pain problem.*

Not surprisingly, these domains of health are all interconnected. Changing one element invariably affects the others. For example, changes in emotions like stress and anxiety (psychology) alter brain chemistry, hormone levels, muscle tension, and immune functioning (biology). Sleep and exercise (biology) fundamentally change our mood (psychology) and ability to work, function, and socialize (sociology). Sociological factors, like environment, lifestyle, diet, and social support, alter our hormones and brain chemistry (biology) as well as our thoughts and emotions (psychology). In fact, these three domains of health—biological, psychological, and sociological—constantly interact in endless feedback loops, each one influencing the others.

While injury and illness clearly matter—and they matter a lot—they simply aren't the full story when it comes to pain. Neuroscience confirms that all information matters to the brain when it's deciding whether or not to make pain, and how much. We'll soon see how this works on a more granular level, peering into the brain and body to get a closer look. But the conclusion here is as obvious as it is profound. Since pain is produced by a combination of biopsychosocial factors, *treatment must be biopsychosocial, too.*

When it comes to chronic pain, it is insufficient to just treat the back, to focus only on the knee. As healthcare providers, we aren't doing our job if we only write prescriptions or just recommend surgeries. And as patients, we do ourselves a massive disservice if we accept this limited, reductionist treatment plan. In order to effectively treat our pain, we must focus on our emotional health, social health, and environmental health, too.

Treatments That Work

I promised good news, and there is more of it.

Not only do we know the truth about pain, but we also know how to treat it.

Over the past decade, we've found that biopsychosocial treatments targeting biological *and* psychosocial factors are the most effective and comprehensive—significantly more so than medications alone.[18,19,20] They also turn out to be more economical, reducing the cost of treatment and shortening our suffering.[21] These interdisciplinary programs include a variety of "nonpharmacological" interventions—frustratingly and unhelpfully defined by the thing that they *aren't*, rather than the thing that they *are*. (Surprising to no one, the long, sticky arms of Big Pharma have polluted even our treatment descriptions.) These treatments sit upon mountains of research, have abundant evidence of effectiveness, and are gaining momentum in medicine.

Unfortunately, many of these biopsychosocial protocols have been marginalized as "alternative" or "complementary," and are consequently plagued by stigma and misunderstanding. To clear up any lingering confusion, these interventions are neither alternative nor complementary: They are simply *medicine*. But despite the evidence, treatments like the ones I'll introduce in this book are often overlooked, omitted, or flat-out rejected. There are many reasons for this, including the erroneous belief that they target only mental health, not physical; a scarcity of trained and available providers; little to no reimbursement by insurance companies (which will gleefully reimburse as many expensive surgeries as you're willing to tolerate, but often not a single psychotherapy session); doctors' reluctance to refer to these treatments due to pervasive cultural skepticism; and rejection by patients, who feel stigmatized, dismissed, and invisible.

Ironically, despite the stigma and misunderstanding that prevent their use, these treatments address the parts of the recipe that are often ignored: the things we think and feel, how often we move, the quality of our diet and sleep, the communities we surround ourselves with, even our trauma histories. Not surprisingly, the very same factors that interact to make pain also contribute to healing it. Targeting these factors, in turn, change our biology—*not just our psychology*—to ultimately change the pain we feel.

Confirmation at the Top

Neither the pain crisis nor the opioid epidemic will change until pain education and clinical practice catch up with science. Until we start expanding treatments to include emotional, social, and environmental health, we will remain stuck. The evidence for this revised approach is now so overwhelming, so universally accepted, that governing bodies and policymakers at the highest levels are clamoring for change—all pushing for a move toward biopsychosocial approaches, and away from surgeries and medications alone.[ii] Despite the fact that it's rarely implemented, the updated, now-standard recommendation for chronic pain is to apply effective nonpharmacological treatments *first*, before potentially addictive medications.[22]

As it is said in medicine: first, do no harm.

But pain medicine isn't changing where it matters most: in our homes, hospitals, doctors' offices, and pain clinics. As long as Big Pharma has a multibillion-dollar budget with which to prey on our ignorance and sell

---

ii Governing bodies advocating for biopsychosocial approaches to pain include the National Institutes of Health, Centers for Disease Control, International Association for the Study of Pain (the largest and most influential pain science organization in the world), Institute of Medicine, Association of American Medical Colleges, Joint Commission for hospitals, and many others.

us the idea that pain is purely physical, we'll continue trying to solve pain with a pill.

Whether you've come to this book because you're tired of treatments that don't work, because you fear your pain has no end, because you're a provider experiencing burnout, or simply because you inhabit an aging body, welcome. We are about to turn pain medicine on its head.

Regardless of what you may have been told—that chronic pain is "untreatable"; that there's only one solution, and it comes in the form of a pill; that pain has no known cause and therefore no known cure—there is hope. Healing is possible, and this book will provide a clear road map. *Tell Me Where It Hurts* is rooted in the most recent neuroscience, evidence, and research. It approaches pain from a completely unique perspective, describing the science of how our emotions, perceptions, and environment impact the pain we feel just as much as any injury. It describes the role of anxiety, depression, and trauma in pain; outlines where our current treatments fail; and offers an optimistic path forward. It bridges the gap between medicine and psychology, physical and emotional, to get to the heart of understanding and treating pain.

Because understanding pain—really understanding it—changes lives. It changed Sam's, and it has changed mine.

I hope it will also change yours.

2

The Real Science of Pain

HOW PAIN WORKS AND WHY WE HAVE IT

The Fastest Man Alive

Chances are high you've heard of Usain Bolt, otherwise known as "Lightning Bolt."

He is the fastest man alive.

A sprinter, Bolt astonished stadiums of stunned spectators when he broke multiple world records at the 2008 and 2016 Olympics. His body is an exquisite, finely tuned running machine. Bolt doesn't just run: He *flies*. Jaw clenched, sweat dripping, his feet barely seem to touch the ground. His speed is not an illusion: Bolt currently holds the world record in the 100-meter, 200-meter, and 4×100-meter races. He has earned nine Olympic Gold medals. He is known to be one of the most naturally gifted athletes on the planet, and the greatest sprinter of all time.

Usain Bolt also lives with a serious health condition. It is called scoliosis.

Scoliosis, a progressive condition in which a patient's spine inexplicably, alarmingly curves, has been known to result in impaired movement, compressed organs, difficulty breathing, and debilitating chronic pain. At its most severe, it can be life-threatening. Scoliosis is typically corrected with a back brace, surgery, or both, the pain managed with medications.

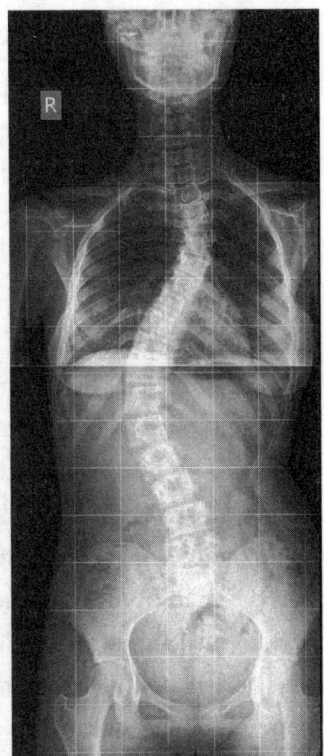

(L) Usain Bolt, fastest man alive.
Credit: Kyodo news via Getty Images
(R) Scoliosis X-ray.
Credit: iStock.com/oceandigital

Shockingly, Bolt's spine is so twisted that his body is asymmetrical—his right leg is half an inch shorter than his left. Researchers who have studied Bolt's seemingly impossible biomechanics have determined that, as a result of his spinal pathology, his left leg remains on the ground 14 percent longer than his right when he runs, and strikes the ground with a completely different degree of force.

One might reasonably think, given the degree of structural damage to Bolt's spine, the unevenness of his hips and legs, and the way these affect his stride, that he would be in tremendous pain. Indeed, this is the lie we've been sold: that damage and pain are one and the same. In fact, had young Usain seen a different doctor, he may have been told that he'd never walk normally...let alone run.

However, this elite athlete instead reports that his back "doesn't really bother him," commenting that he doesn't have much pain.[1] In fact, Bolt's spinal issues have impeded him so little that they remain uncorrected.[2] Even more astonishing, doctors believe that, rather than encumbering him, Bolt's abnormal gait and asymmetry are the very things that help him run *faster*.[3]

It is, of course, true that bones can break, muscles sprain, joints swell, and tendons tear. It is true that spines can twist and warp. And it's true that these things require attention and care. But while body parts may get damaged, *that's not where PAIN lives.*

If the myths we've been sold about pain were true—if tissue damage and the anatomical abnormalities found on back scans were the sole source of pain—Usain never would've been able to streak across our screens like a blistering bolt of lightning. He never would've stood atop an Olympic podium, powerful muscles shining, adorned with medals and a Jamaican flag, proudly waving to admiring crowds. Instead, he would be immobilized, disabled by pain, relegated to a hospital bed—watching the Olympic Games on TV, like the rest of us mere mortals.

It's time to scrap everything you've been told—to rewrite every myth you've been sold. Because pain is so much more than what we've been led to believe. To have any chance of effectively treating it, we must peel back its layers and examine its nuances.

To do that, let's begin at the beginning: with a visit to your doctor.

The Mismeasure of Pain

If you've ever been to a doctor's office or hospital, you're likely familiar with our current method of assessing pain. It looks something like this:

A familiar pain scale.
Credit: Illustration of Wong-Baker Scale® with emoji by Belbury, CC BY-SA 4.0 https://creativecommons.org/licenses/by-sa/4.0, via Wikimedia Commons, based on the original scale created by Donna Wong and Connie Baker.

This illustration is typically accompanied by a question: "On a scale of 0 to 10, how much pain are you in?"

Some have suggested that this scale is inadequate and invalidating, epically failing to capture the pain experience. As someone who confronts this scale daily, I find it problematic, too. Here's my interpretation of what these faces actually seem to convey:[i,4]

0. I'm so excited that it's summer vacation!
2. I like going to the zoo.
4. What do you mean you ate all the popcorn?
6. I can't believe it's going to rain on my birthday; that's so disappointing.
8. My favorite show was canceled? How terrible.
10. The end of the movie *Titanic* is so sad!

---

i Credit to author Allie Brosh for inspiring this pain scale interpretation from her *Hyperbole and a Half* blog post, "Boyfriend Doesn't Have Ebola. Probably." I regularly prescribe this article to my patients, and now prescribe it to you.

Emojis aside, this basic scale begs some basic questions. For example, what does a 10 on the pain scale actually *mean*? Is it being bitten by a shark, a root canal, pushing an eight-pound baby out of a five-inch birth canal? What's the difference between a 9 and a 10 anyway, does anyone really know…? Why is my broken leg a 9 and yours a 5? It generates more nuanced questions, too: Does a 4 accurately represent my suffering if I've had pain for twelve years across twelve body parts, and how does that compare to a brief 10 in one body part? If my 8 temporarily drops to a 4 when you distract me and make me laugh, does it require any less attention?

Science agrees with our skepticism. Pain is never composed of a single quality that varies only by intensity. Nor is it fixed, objective, or even linear. While it's useful to gather this data about real-time pain intensity, pain is subjective and contextual, constantly changing as our mood, environment, and bodies change. Neither are these ratings correlated with, nor predictive of, how well we're able to function or what we're able to accomplish. Take two people with an identical pain rating, and one might be game to wrangle a steer while the other might be couch-bound. And because pain is relative, there's no one correct way to assign—let alone interpret—these numbers. In fact, there is *no such thing* as an "objective" measure of pain.

But we don't just get pain assessments wrong. Perhaps the reason we fail to accurately measure pain is because we've failed to meaningfully, and accurately, define it.

The first question I ask my patients—people who have been in pain for years without relief—is: "Has anyone ever explained pain to you?"

To date, not a single person has ever said yes.

Let's end that now.

What You Don't Know *Will* Hurt You: Pain Science 101

Despite the fact that pain is a ubiquitous human experience, affecting all of us at some point in our lives, basic definitions of pain are astonishingly inconsistent, changing depending upon whom you ask. Your physical therapist is likely to give you one answer; your physician another; your psychologist another; your mother, another. This is true even *within* medicine, driving the different solutions we prescribe: A surgeon evaluating an abnormal back scan may believe pain is due to a structural abnormality, something to be excised with a scalpel. An anesthesiologist may prescribe medications to solve your pain problem. A chiropractor is likely to blame misalignment and recommend an adjustment. And a psychotherapist is prone to focus exclusively on emotional pain. As with most things, the truth lies somewhere in the middle: All factors contribute. This is why an accurate, universal definition of pain is so critical—our treatment recommendations are contingent upon our understanding of the issue at hand.

The official definition of pain hasn't evolved much over the past forty-five years. This definition, established by the International Association for the Study of Pain, the leading global authority on the study and treatment of pain, is "an unpleasant sensory and emotional experience associated with, or resembling that associated with, actual or potential tissue damage."[ii,5] This definition is critically important, because it reveals that while pain is in part defined by sensory characteristics—for example, location, duration, quality, intensity—it is also always emotional.

And then, it is much more.

Pain is our body's warning system, a process triggered by actual or potential danger. It can be triggered by an injury, an illness, or—as we will soon see—our bodies being out of balance. Perhaps most importantly, pain is an *interpretation*—our brain's best guesstimate based on all available information as to what's happening and what's to come, a predictive

---

ii The definition of pain has been updated only once since 1979, in 2018, and not by much. *Sensory* here refers to the spatial and temporal characteristics of the pain experience, as well as its quality—burning, prickling, throbbing—and is often referred to simply as intensity. The emotional or affective dimension of pain refers to how unpleasant, aversive, or "bad" the pain is. It is both of these, together, that make pain *pain*: neither is sufficient on its own.

process meant to save our lives. An evolutionary response to perceived threat, pain is protective: an alarm that grabs our attention, motivates us to change our behavior, and buys us time to heal.

Pain is also a teacher, training our brain to learn from past mistakes and avoid dangerous situations in the future. Pain is the reason we pull our hand from that hot stove, stop running on a twisted ankle, and go to the dentist to get that rotting tooth pulled. Simply put, pain saves our lives. People born with a dysfunctional pain system and feel little pain, despite how luxurious that may sound, don't live very long.

But not all pain is created equal. Temporary, short-term pain, called acute pain, is the type we're most familiar with. Acute pain is typically the result of an injury or illness, and is defined as pain in any body part lasting three months or less. Common examples are a skinned knee, a broken bone, the pain of childbirth, and muscle pain from the flu. Acute pain also describes new, unexpected pain: a sudden stomachache after eating spoiled food, the abrupt pain of muscle fibers torn during a workout, the full-body pain of a spiking fever. The quality of acute pain is often as important, and as informative, as its location. For example, intense, stabbing abdominal pain can indicate a ruptured appendix, while tingling, electric pain can indicate nerve damage, and red-hot, throbbing pain can indicate an infection. In acute, emergent situations like these, it's important to pay attention to these danger messages—to stop activity, seek help, and rest. This lifesaving instinct gives our bodies an opportunity to heal.

If pain persists for three months or longer, or beyond expected healing time, it is called chronic pain.[iii] This time stamp is arbitrary; in truth, chronic pain has no agreed-upon definition beyond "pain that endures." Like acute pain, chronic pain can appear anywhere in the body—head, pelvis, back, jaw. Unlike acute pain, chronic pain does *not* serve an

---

[iii] This discrete time stamp is widely debated. The transition from acute pain to chronic pain is more likely a gradual process that occurs over time, evolving differently in different people, rather than morphing from one to the other at *exactly* the three-month mark.

adaptive purpose. Rather, it is a sign of a signaling system gone awry, an alarm sounding in the absence of danger. Like every system in the human body, the pain system can malfunction, too. Rather than facilitating survival, chronic pain can instead interfere with life and functioning. In fact, as we will soon see, chronic pain has less to do with the body part that hurts and more to do with changes to our brain and nervous system.

But chronic pain isn't simply pain that lasts longer. Neuroscience suggests that chronic pain actually works differently, involving biological mechanisms that are distinct from acute pain. Because of this, chronic pain is considered a disease process in its own right. Chronic pain can be linked to a specific condition, such as migraine, sickle cell disease, or cancer, or it may have no known etiology or cause. Pain that isn't connected to any known structural damage or disease is also called idiopathic pain. Interestingly, research suggests that a large proportion of chronic pain is idiopathic[6]—yet another clue that we've been looking for the source of our pain in all the wrong places.

Pain and the Brain: The New Neuroscience of Pain

It's easy to believe that pain lives exclusively in the part that hurts—our bad back, our aching knee. Indeed, this is what we've all been told.

But neuroscience confirms this isn't true. One of the reasons we know this is because of a condition called phantom limb pain, wherein someone loses a limb—an arm or a leg—but continues to feel terrible pain in the missing body part. The fact that we can have terrible leg pain in a leg that is no longer attached to our body tells us that pain can't possibly be constructed by the leg alone. Because if pain lived exclusively in the body, *no limb should mean no pain*. Pain must therefore be constructed somewhere else, and neuroscience reveals that this somewhere else is the brain.

And while we may all learn this eventually, no one knew this better than Mateo.

The Phantom Hand

Mateo, a shy, quiet twelve-year-old, was still in the hospital when his medical team called me. "I've never seen anything like it," said his surgeon. "No amount of pain medication is helping. Can you get him in?"

Six weeks earlier, Mateo's friend had found a firecracker in his parents' garage. He brought it to the park, where the middle schoolers excitedly gathered in a circle. Mateo and his friend lit some matches. The firework had a long fuse, so the children assumed they had ample time to toss it in the air and run before it went off, as they'd seen in cartoons. Instead, it detonated instantly. The blast seared the skin on Mateo's face, blew a hole in his eardrum, and burned his cornea. It also took off his left hand. Bleeding and unconscious, Mateo was airlifted to the nearest hospital. A police officer called his mother to tell her that Mateo might not make it.

Thanks to his amazing medical team, Mateo did make it. But his left hand and forearm were irreparable, and the doctors had to amputate below the elbow. After surgery, the pain was excruciating. Neither opioids nor any other pain medications helped.

Within days of leaving the hospital, Mateo and his parents arrived at my office. Quiet and self-conscious, Mateo was a wisp of a child who wore an oversized sweatshirt that swallowed him whole. Mateo answered all questions with a nod or shake of his head, averting his eyes and sucking nervously on his sleeve. He seemed to want to disappear. His father mercifully stepped in and described how the accident had completely transformed his son.

A previously cheerful sixth grader who loved tree climbing, football, and adventuring, Mateo was finding it difficult to adjust to his injuries. He couldn't bathe himself or open a door, let alone climb a tree. He couldn't stop thinking about the accident or the other kids who'd gotten hurt. He felt guilty, wondering how much of it had been his fault. Mateo was too traumatized to sleep alone, and had reverted to sleeping in his parents' bed. He had terrible nightmares. Some nights, his father disclosed quietly, Mateo even wet the bed. On the few occasions Mateo did leave the house, he wore an oversized sweatshirt with long sleeves to hide his missing arm.

Mateo was also in terrible pain. But it was the strangest kind: The pain emanated from his left hand, the one that had been amputated. When asked to describe it, Mateo said it felt like there was a "ghost hand" still attached to his body, a hand that ached and spasmed. He had no fingers, but sometimes it felt like his fingers were cramping. It was confusing and alarming: How could he possibly have pain in a hand that was no longer there?

I gently explained to Mateo that he was right—he wasn't feeling pain in his hand at all. Instead, I told him, there's a map of our entire body that lives in our brain called the homunculus. When a limb is suddenly lost, it can take time for the homunculus and the rest of the nervous system to update and register the change. As a result, amputees sometimes continue to feel sensations in the missing body part, referred to as a "phantom limb." Some people, for example, report feeling like a missing hand is gesturing, pointing, or even like it's picking something up. Up to 80 percent of people with a phantom limb experience pain in this missing body part, a phenomenon known as phantom pain. Phantom pain is due to a variety of factors, including failure of the brain to recalibrate, damage at the injury site, and a nervous system stuck in alarm mode.

Mateo understood, but the information fell flat. It was just science, and it didn't dim his suffering. As the weeks wore on, Mateo withdrew, dropped out of school, and slept all day. He barely spoke or left the house. The grief, misery, and pain threatened to consume him.

He needed help, and fast. I called a local occupational therapist, a specialist who helps people regain function and independence after an injury. Together, we started Mateo on a course of mirror therapy—a neurorehabilitation technique that relies on the power of neuroplasticity to remap the brain and reduce pain. In this treatment, a mirror is placed between a patient's healthy limb and the affected one such that they cannot see their affected limb. The mirror reflects the healthy limb, creating a visual illusion of two strong, functioning arms: one real, the other a reflection. Patients then engage in movement-and-touch exercises while watching the reflection of the healthy limb in the mirror. This illusion feeds visual, motor, and sensory data back to the brain, which perceives the amputated limb behaving normally. With time and repetition, these new neural inputs stimulate neuroplasticity—helping the brain update with new information, and recognize that danger messages are no longer needed.

For Mateo, watching his missing left hand—which was really just a reflection of his right—flex and pick up objects was as satisfying and empowering as it was bizarre. The "muscle spasms" in his phantom hand disappeared within weeks. After three months of treatment with me and his occupational therapist, the pain began to gradually abate, too. That a treatment targeting his brain could actually heal pain was a revelation for Mateo and his family. We celebrated his remarkable progress.

But our joy was premature. Something was wrong. While Mateo's pain was less, he continued to be despondent and withdrawn, a shadow of the vibrant child he'd been before. He was barely eating, and wouldn't go to school. Mateo clearly needed more than just mirrors: He needed an antidote to the guilt and the shame. He needed to know he wasn't alone. He needed to know he could still live a full life, that his dreams hadn't died. This would require hope, motivation, and resilience in large quantities. It was an extremely tall order.

Luckily, I knew just the guy.

Emmett the Explorer, a famous archaeologist, was an old friend from graduate school who excavated massive dinosaur bones from forgotten caves around the world. He'd become a global phenomenon when a local TV network picked up his adventures. But four years ago, after a terrible accident on a dig site, one of Emmett's legs had to be amputated. Like Mateo, Emmett had initially experienced terrible phantom pain. Upon meeting with medical experts, he discovered that targeting his brain—not just his leg—was key to his relief. Emmett sought out various ways to heal, including mirror therapy. It took months of hard work, but in time, Emmett's phantom pain slowly receded. Today, he is again exploring, adventuring, and enjoying a rich, full life.

Mateo loved adventure shows, and had even watched episodes of Emmett's series filmed prior to Emmett's accident. What Mateo didn't know was that Emmett was now also an amputee. When I told him about Emmett's lost limb, phantom pain, and subsequent adventures, Mateo was stunned. Over the next few days, he binged all of Emmett's new episodes, watching wide-eyed as Emmett rappelled out of helicopters into overgrown jungles with one leg. He watched as Emmett, unafraid, climbed, swam, and hiked using a robotic limb. He watched as Emmett unashamedly pulled off his prosthesis, in full view of camera and crew, revealing his stump. In session, Mateo excitedly recounted Emmett's adventures. It was forward momentum, and I wanted to harness it.

I called Emmett and asked for a favor. He agreed to meet with Mateo and me the following week via video. On the day of the meeting, Mateo was visibly nervous. But when Emmett's familiar face flashed across the screen, Mateo's eyes filled with delight. Brimming with confidence, Emmett held up his stump for Mateo to see. Mateo shyly rolled up his sleeve and showed Emmett his. Emmett told Mateo about the phantom leg pain he'd experienced, and how it had gone away. He proudly showed Mateo his prosthetic, robotic leg, detailing all the things he could do with it. Mateo would love having a robotic arm, Emmett said enthusiastically, explaining that he could use it for everything from football to fishing. "Watch out world," boomed Emmett with a conspiratorial grin. "We'll both be bionic!"

After that, things started changing—and fast. After months of resistance, Mateo finally agreed to get fitted for a prosthetic arm. I'd been teaching him strategies to manage pain and gain control of his body, and he was using them. When asked, Mateo said that meeting Emmett, following his pain protocol, and coming to treatment were transformative. As Mateo's physical, emotional, and social health improved, I told him, it was very likely that his brain—and his pain—would continue improving, too.

I was working in my office a few weeks later when Mateo's father called to say there'd been a visible change in his son. After months of precipitous weight loss, Mateo had finally started eating again. He was using his prosthetic arm to dress and open doors, and had asked his father to help him climb a tree. He was brighter, happier. Mateo was finally returning to himself and the activities he used to love. He'd even asked to go back to school. His father's voice crested with joy. Mateo still had a long journey ahead, but his improvement was palpable. "There's been another big change, too," his father said. "But I can't tell you. You'll have to see for yourself."

When Mateo returned to session the following week, his oversized sweatshirt was gone. Instead, he wore a plain T-shirt, his arms and remaining hand in full view. I complimented his bravery and asked what had changed. Mateo sat tucked into the couch—small, quiet. He thought for a moment, then pulled himself up to his full height and said: "Well, if Emmett isn't afraid to show his leg, I'm not afraid to show my hand."

His face shone with hope and promise. His incredulous parents later remarked that Mateo had gone from a child who believed he could do nothing…to a child who believed he could do *anything*.

* * *

Mateo's story opens the door to the fascinating neuroscience of pain. As we begin our journey, the first thing to know is this: There is no one single "pain center" in the brain. Rather, pain is a diffuse neurological process constructed by multiple parts of our nervous system. This includes the prefrontal cortex, the foremost part of the brain, which helps direct attention, make decisions, and assemble predictions; the somatosensory cortex, which processes sensory information from the body like touch, pressure, and proprioception (the location of your body in space); an area of the midbrain called the periaqueductal gray, a structure involved in stress and pain modulation located deep in the center of our brain; and a collection of brain sites historically known as the limbic system, a critical part of our emotion machinery. Limbic structures include the amygdala, famous for its role in fear, threat, and emotion processing, and the hippocampus, which stores memories of past pain experiences and connects these to emotions.[7,8] An important gateway, the spinal cord acts like a bouncer, gating and regulating sensory information from our tissues up to the brain—called bottom-up pain processing—and from the brain back down to the periphery, known as top-down processing.

PARTS OF THE BRAIN THAT MAKE PAIN

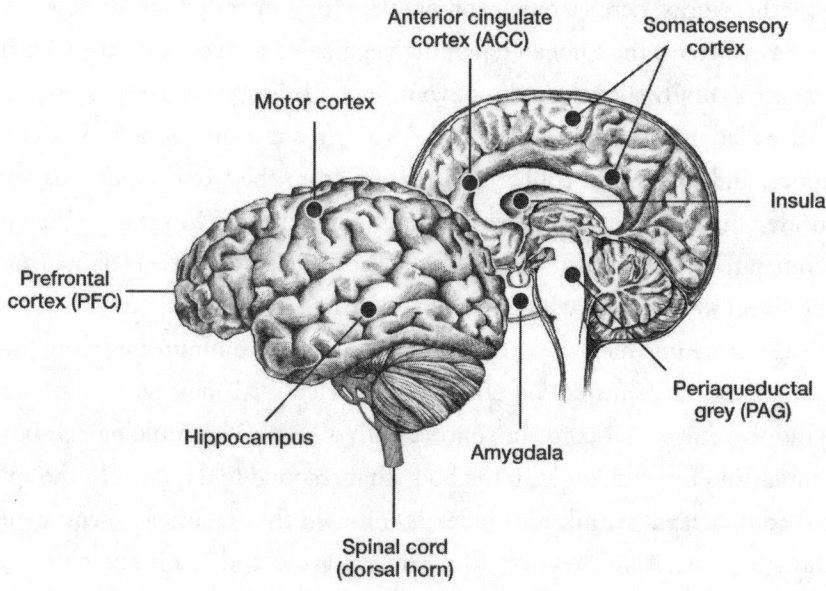

Credit: Illustration by Eva Huzella

At its most basic, the human nervous system consists of two interconnected parts: the central nervous system (CNS)—brain and spinal cord—and the peripheral nervous system (PNS), the network of nerves running through the rest of our body, which connects everything from our toes to our nose. Together, the CNS and PNS collect and interpret all available information from inside us and around us. The brain, our body's prediction machine, then decides how to respond based on all available data: Is there an emergency, or is this a false alarm? If there is danger, how significant is it? What behaviors need to change in order for us to stay safe?

This data matters a lot to our pain system, our body's danger-detection system. If the brain assesses all of this collective data and determines our body is in danger, *it makes pain to protect us*. Our appraisal of the situation—whether we believe we're in danger or not—is a critical determinant of the pain we feel. Because it's useful to understand the basic biomechanics of how this works, here's a quick and painless (pun intended!) breakdown.

Say you're in the kitchen chopping vegetables with a sharp knife when you accidentally slice the palm of your hand. This activates sensory receptors called nociceptors: specialized cells in our skin, organs, muscles, bones, and joints that collect information from the world around us and inside of us. Nociceptors detect changes and extremes in pressure, touch, temperature, chemicals, and tension.[iv] These cells derive their name from the Latin word *nocere*, which means "to harm."

The warning messages they generate, which communicate potential danger, are transmitted via chemical and electrical messages from your hand to your spinal cord, in concert with a host of immunological, hormonal, and other changes in the body that respond to damage. In the spinal cord, a rapid withdrawal reflex is initiated that requires no conscious thought or decision-making. This helps us avoid and minimize potential damage: For example, you quickly and instinctively drop the knife and pull your hand away. This protective mechanism is adaptive, saving time in an emergency—imagine if you had to sit down and filter through all your options in the face of every imminent danger? We might bleed to death while we pondered.

This process of detecting and reflexively responding to potential damage is called nociception. Nearly all animals, even those with simple nervous systems, including jellyfish and fruit flies, experience it. Nociception

---

iv A common mistake people make is referring to nociceptors as "pain receptors." *This is incorrect.* There is no pain until sensory data reaches the brain and is interpreted as such. In the words of renowned pain experts Ronald Melzack and Joel Katz, "Pain processes do not begin with the stimulation of receptors. Rather, injury or disease produces neural signals that enter an active nervous system that (in the adult organism) is the substrate of past experience, culture, and a host of other environmental and personal factors. These brain processes actively participate in the selection, abstraction, and synthesis of information from the total sensory input. Pain is not simply the end product of a linear sensory transmission system; it is a dynamic process that involves continuous interactions among complex ascending and descending systems." Source: Melzack, R., & Katz, J. (2013). Pain. *Wiley Interdisciplinary Reviews: Cognitive Science*, *4*, 1–15.

is sometimes accompanied by automatic, autonomic bodily responses like sweating, rapid heart rate, changes in blood pressure, lightheadedness, and even fainting.

However, this experience is *not yet pain*. While it's possible to experience tissue damage without awareness, there's only one organ in the human body that enables us to consciously perceive and interpret sensations and experiences: our brain. It's therefore only when these messages reach the brain that they become the experience we know as pain.[9] The brain evaluates data from our environment (cooking in the kitchen), prior knowledge (sharp knives are dangerous), memories of past experiences (previous injuries from sharp objects), input gathered via the five senses (lacerated skin, blood on the counter), and emotions (panic!), among other things. Multiple brain sites now work together to reach a conclusion about what's happening and how to respond. For example, rather than continuing to cook, we might get bandages, call for help, or go to the hospital. The brain also sends information and instructions back down to the body via the spinal cord, accompanied by a host of immunological, musculoskeletal, chemical, hormonal, and nervous system changes.

Indeed, a story like this one—a story about blood, breaks, and bruises—is a classic example of what we think of when we think of pain. But the truth is that this barely scratches the surface. If pain were a simple matter of skin and senses, someone with a back like Bolt's would never reach superhuman speeds, or achieve impossible feats of strength.

Like all good stories, there's much more to this one.

Hurt and Harm Are Not the Same

Most of us believe that pain is caused by an old running injury, a herniated disc, or years of terrible posture. However, while these factors can contribute to pain, *they aren't actually the cause*. That's not to say that damage can't cause pain: As we just saw in the kitchen, it most assuredly can, and decidedly does. It just doesn't fully explain the pain we feel, nor is it ever the sole driver.

This revelation is at the heart of understanding the true science of pain, and opens the door to other crucial truths, too. One of the most important is that "hurt," or pain, and "harm," or physical damage, are not the same.[10] They commonly co-occur—but they don't always. Pain can occur in the absence of damage, and damage can occur in the absence of pain.

We don't need an Olympian to demonstrate this point, although Bolt handily helps. We've experienced it ourselves many times. Those black-and-blue marks you discovered in the shower, but had no idea how they got there? Evidence of crushed capillaries and blood vessels, without accompanying pain. And that time you played a stunning basketball game, only to discover afterward that your leg was covered in blood…? A perfect example of harm, or damage, minus the hurt. This is a common experience for athletes, who regularly report discovering blood and bruises only after the game is over—even though the injury occurred hours before. Cognitive, emotional, and contextual factors like excitement, distraction, and lack of conscious awareness can keep pain at bay. And the opposite can also occur, as you'll soon see: hurt without harm.

What explains this strange disconnect?

The Meaning of the Message: Context Matters

Across cultures, certain painful experiences are considered normal—even enjoyable. Consider the ones you've deliberately sought out in your lifetime: tattoos and painful piercings; eating spicy, hot foods until our mouths burn and our eyes water; childbirth, which we do again and again, no matter how painful; grueling workouts that cause days of soreness; contact sports like boxing and martial arts; 26-mile marathons; and rough play during sex, like spanking or biting. The titillating subject of bestselling books like *Fifty Shades of Grey*, BDSM—which stands for bondage, dominance/discipline, sadism/submission, and masochism—isn't technically "freaky" or uncommon, but is rather reportedly enjoyed in some form or another by one in three Americans.[11]

We seek out emotional pain, too: the thrill of scary movies, the terror of gravity-defying roller coasters, and that time you watched *The Notebook* simply because you needed a "good cry." Not to be ignored, there are also those who engage in self-harming behaviors, like cutting, because it provides, for them, physical and emotional relief.

Are people who enjoy muscle soreness, spicy food, and spicy sex abnormal or genetically flawed? The answer is no—not even a little. Pain may be biological, but it is also *contextual*. This is why an experience that feels painful and miserable to you can feel deliciously gratifying to someone else, even when the sensory input—and tissue damage—are exactly the same. And an experience that may feel unpleasant in one context can feel good in another, even to you, even on the same day. For example, a rough massage from a trained therapist in a serene environment can feel good, while those very same sensations might feel agonizing during an assault. And a squeeze on the neck during sex with a trusted partner might feel exciting and erotic, while that same injury sustained during a mugging might feel violent, and violating.

Because pain is a guesstimate—our brain's best assessment of whether our body requires protection—context is critically important.[12,13] "Context" refers to the sum total of all available information from our internal and external environment from which our brain makes meaning and predictions in any given moment. External contextual cues include where we are, who we're with, and what's happening around us. Internal contextual data from our bodies include expectations and beliefs, the *meaning* we assign to sensations, how we're feeling emotionally, and prior knowledge. These internal inputs, in turn, then influence how we perceive and react to the external world—further shaping our experiences.

The impact of context on pain is all around us, punctuating our everyday lives. For example, if you stub your toe on the day you get fired, it will feel substantially worse than that very same injury sustained on a relaxing Sunday at the beach with friends. Memories of a past debilitating concussion can exacerbate the pain of a new and different head injury. People and places matter, too: An arm pinch feels worse if made by a frighteningly long needle than by a friend's reassuring squeeze. Pain hurts more when we're lonely and sad than when we're happy and surrounded by people we love. Pleasant colors and safe, soothing surroundings can lower pain volume, which is why children's hospitals are festooned with murals and stuffed animals. The color red, known to signal danger in nature, has been shown to amplify pain intensity, while calming colors can make pain feel less bad. Interpretations change pain, too: unexpected chest pain feels worse if we believe it's due to a heart attack than to a basic cold. If our doctor appears professional, intelligent, and trustworthy, our pain is likely to feel less bad than with a provider who seems less so. Negative diagnoses and poor prognoses can actually worsen symptoms and amplify pain. And if we take a sugar pill—a placebo—believing it's an effective pain medication, our pain can subside simply because we expect it to.

Sensory input doesn't exist in a vacuum: Social and environmental contexts change the meaning of the message. This mechanism helps us make sense of the constant input bombarding our brains, teasing out which sensations are safe—and thus don't require significant attention or resources—and which are potentially dangerous, requiring an alarm response. This is why, despite the same degree of harm, we can experience a different degree of hurt: *The meaning we make of the message matters.*

To illustrate this point, I give you A Tale of Two Nails.

A Tale of Two Nails

In 1996, the *British Medical Journal* reported that a twenty-nine-year-old construction worker had been working on a job site when he accidentally jumped off a plank—straight onto a seven-inch nail.[14] Much to his horror, it drove straight through his boot clear to the other side.

In terrible pain, the man was rushed to the emergency room. "The smallest movement of the nail was painful," wrote his doctors, so "he was sedated with fentanyl and midazolam"—a powerful opioid painkiller and a sedative.[15] When his doctors removed his boot, they discovered that a miracle had occurred: The nail had passed through the space between his toes. There was no puncture wound, no blood, no tissue damage—not a

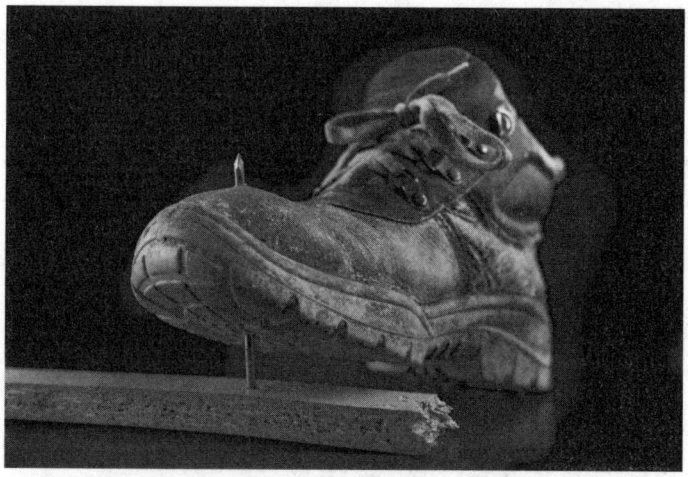

Credit: iStock.com/Piotr Wytrazek

scratch to be found. But despite the absence of injury, his pain was real. How could this be possible?

As our brains are designed to do, his brain used all available information to make meaning of the message—to determine whether or not he was in danger, and how much. This included knowledge about his hazardous work environment; memories of past pain experiences; beliefs and predictions; input from his five senses, including the visual of a giant nail sticking out of his shoe; his coworkers' horrified faces; emotions, including fear; and other data. His brain then made a guesstimate about what had happened and how to respond. Because it *perceived* potential danger, it made pain to protect him.

The second Tale of Nails takes place ten years later in the ski resort town of Breckenridge, Colorado. Patrick, also a construction worker, was working on a job site when his nail gun misfired. It recoiled backward, the butt of the gun clocking him in the jaw. Patrick saw a nail shoot across the room and bury in the wall across from him. Other than a vague toothache, mild headache, and some swelling, he believed he'd escaped unscathed. Six days later—six days of eating, sleeping, and working—Patrick decided to go to the dentist to see about that tooth. The doctor took a scan of his patient's jaw. Much to both men's surprise, they discovered a four-inch nail embedded in Patrick's face. It had plunged 1.5 inches into his brain, barely missing his right eye. Removal of the nail required a four-hour surgery. Said the surgeon afterward: "He's the luckiest guy, ever."[16]

In this case, there was significant damage—but very little pain. How could this be?

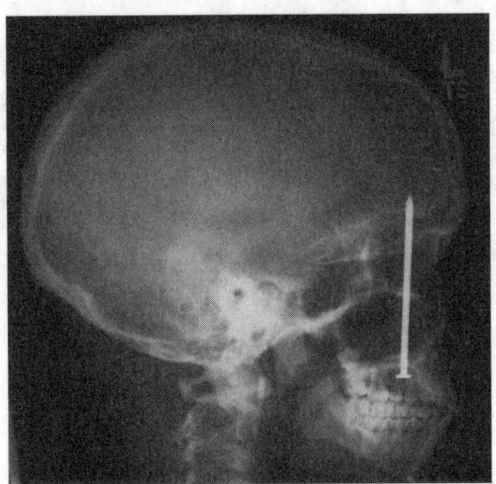

Credit: Photo courtesy of Dr. Seth Reiner

As with our other construction worker, Patrick's brain also used all available information to assess for danger and make meaning of the message. But in this case, Patrick saw a nail travel in the opposite direction of his body—contextual information that convinced his brain that risk of danger was low. Because the available physical, emotional, and environmental cues failed to trip his danger alarm, Patrick's pain system remained relatively quiet—despite serious bodily harm.

These stories remind us of an important fact, one easily forgotten when something hurts: *Pain is not an accurate indicator of tissue damage.* The amount of pain we feel doesn't reflect the degree of damage to our bodies. We can experience significant pain without any damage—for example, a nail to the boot but not the foot—and very little pain in the presence of significant damage, like a four-inch nail to the face.

Harm, but No Hurt: Wrinkles on the Inside

This surprising disconnect between pain (hurt) and damage (harm) isn't just reserved for tales of nails. It's evidenced by even the most common conditions, like low back pain, which will affect more than 80 percent of us over the course of our lifetime. Despite the fact that back pain is routinely blamed on slipped discs, herniations, and other structural issues found on back scans, studies shockingly reveal little to no correlation between these "abnormalities" and pain.[v,17,18] In fact, these abnormalities

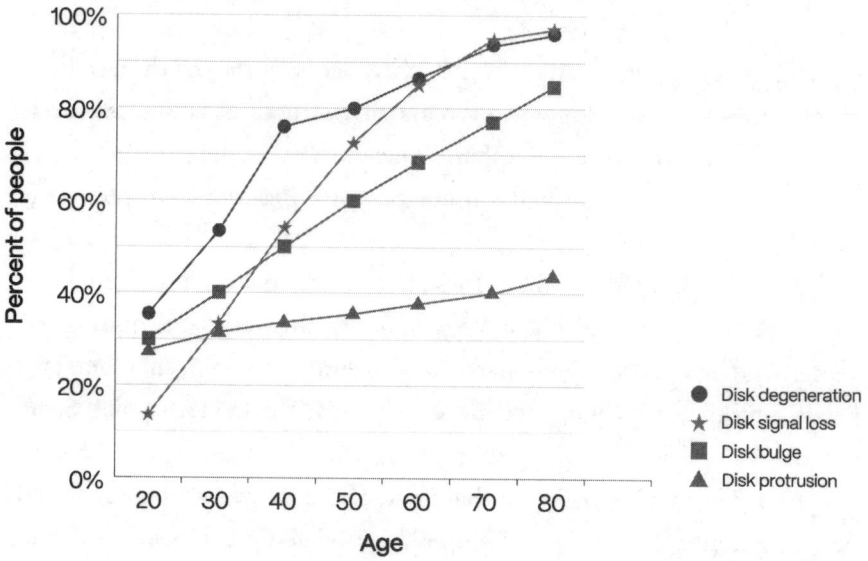

Physical damage doesn't always indicate pain, and pain doesn't always indicate damage.

Credit: Illustration by Eva Huzella. Graph based on data from Brinjikji, Luetmer, Comstock, et al. (2015). Systematic literature review of imaging features of spinal degeneration in asymptomatic populations. American Journal of Neuroradiology, 36(4), 811–816.

---

v   Lest you should think this rare, research shows that 85 percent of low back pain has no known structural or biological cause—telling us, once again, that damage and back pain are *not* one and the same. Source: Koch, C., & Hänsel, F. (2019). Nonspecific low back pain and postural control during quiet standing—A systematic review. *Frontiers in Psychology, 10,* 586.

don't seem to predict pain at all. When scientists scanned the backs of over three thousand healthy people with zero back pain to peek inside, they found these very same abnormalities. Nearly all of the individuals scanned—90 percent of subjects ages sixty to sixty-nine, and 80 percent of subjects ages fifty to fifty-nine—had bulging discs, disc degeneration, and other anatomical abnormalities—but *no accompanying pain*.[19]

In yet another study of more than 1,200 healthy subjects who had no pain, nearly 90 percent of them had bulging discs.[20] No hurt; measurable harm. If hurt and harm were the same, individuals with abnormal scans should be in terrible pain—and the more abnormal their scan, the more pain they should feel. But that's not what scientists have found at all.

If this comes as a surprise, it's only because the truth has been obscured. Science has known for years that structural issues are rarely the cause of low back pain—less than 5 percent of cases, in fact—and therefore, that surgery, spinal cord stimulators, and injections are rarely the cure.[21]

Lest we're tempted to believe these results are an error, an anomaly, or something just to do with the biology of backs, this very same finding has been found across other body parts, too, including the hip, shoulder, jaw, pelvis, uterus, wrist, knee, and neck.[22] Indeed, the evidence that tissue pathology doesn't sufficiently explain pain is utterly overwhelming. Scientists have reasonably concluded that most of the lumps and bumps found on scans are most likely due to normal, age-related processes rather than anything sinister or pathological. Developing "wrinkles" on the inside—on our spine as well as on our face—is simply a part of the normal aging process. A dermatologist wouldn't tell you that your face is "degenerating." Why, then, would we say this about your back?

This doesn't mean we should ignore growths or broken body parts, or that these don't contribute to pain. We shouldn't, and they do—particularly in the case of acute, short-term conditions. But study after study reveals that injuries, deformities, and degeneration are found in just as many people who have pain as those who *don't*. Millions of us are currently walking around, quite happily and ignorantly, with bulging discs, crooked body parts, misalignments, and more—and we don't know it or don't mind, because we have no pain.

There's just no denying the data: Damage isn't always the cause of pain, and it's never the only culprit.

Pain, it turns out, is much more interesting than that.

Pain Is Biopsychosocial: The Three Pillars of Pain

Here's what we do know: Science reveals that pain is biopsychosocial, produced by a combination of biological, psychological (emotional, cognitive, behavioral), and sociological (social, environmental, contextual) factors that work together to create the pain we feel. Pain lives in the glorious, messy middle of all the things that make you, you.

While holistic healthcare and mind-body medicine have tried to change the narrative for decades, even they miss the big picture when it comes to pain. Because understanding pain isn't just about connecting mind and body. The things going on around us—not just inside of us—change pain, too.

In order to treat pain, therefore, we can't just fixate on the part that hurts; we must instead treat *all of you*.

Pain Is Biological

The first pillar, the biological domain of pain, includes the ingredients we've heard the most about: genetics, tissue damage, system dysfunction, inflammation, the wear and tear of an aging body, neurotransmitters, hormones, diet, sleep, exercise. Anything and everything that can go wrong with the human body lives here: blights and diseases, afflictions and infections, things that break, tear, swell, and pull. For Sam, the teen with crippling pain that kept him in bed for four years, this included a

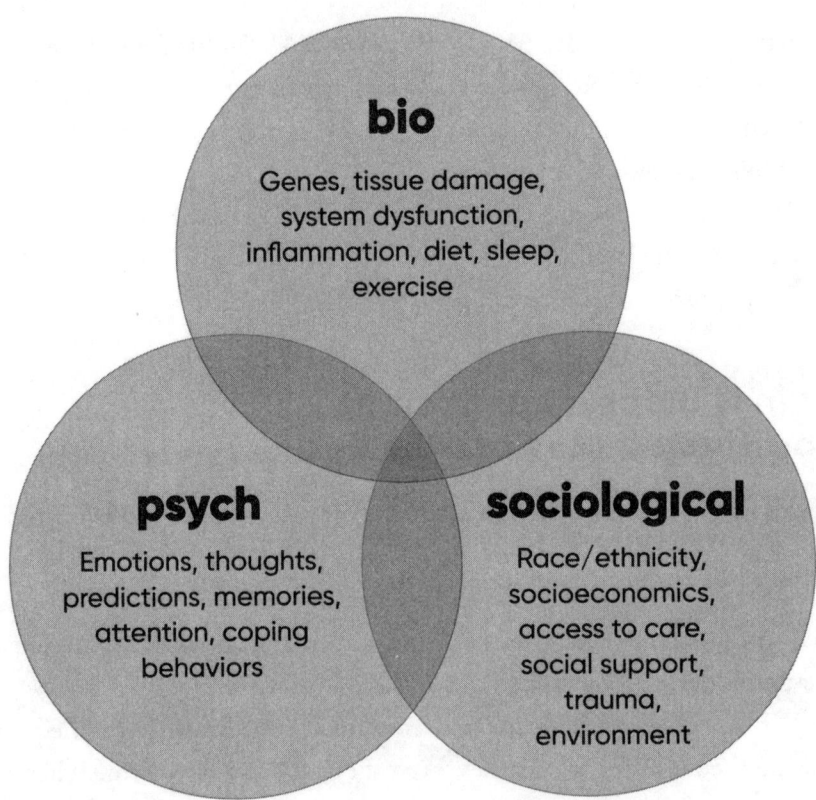

Pain is biopsychosocial, always.

family history of migraine, poor sleep, inadequate nutrition, changes in hormones and brain chemicals, eyestrain, and inflammation. For Mateo, biological contributors included a debilitating injury, a confused brain map, and a pain system stuck in danger mode.

Biological factors are critically important when it comes to pain, and cannot be ignored. They are commonly treated with medical interventions like medications and surgeries—which, while extremely helpful for acute pain, often saving limbs and lives—are considerably less helpful for chronic pain.

Of note, just because pain has a biological component doesn't mean it's genetic destiny or assigned at birth—that is, just because Dad suffered from sciatica doesn't necessarily mean we will. In fact, there's no one single "pain gene" passed down from generation to generation. Rather, scientists believe that hundreds of genes contribute to the pain experience. Some of these genes help determine our "pain threshold," or how much input we can tolerate until it hurts.

But even if the genetic deck is stacked against us, humans are never the product of genes alone. Rather, we are always a combination of our genes (nature) *and* our environment (nurture). So while some painful conditions are heritable to some degree, 100 percent of them, from arthritis to sickle cell disease, are massively influenced by social and environmental factors: things like diet, culture, socioeconomic status, trauma, and stress. Which brings us to the next, and most misunderstood, pillar of pain: the psychological domain.

Pain Is Psychological

The psychological domain of pain is less commonly addressed due to pervasive, and understandable, stigma—as we saw with Sam. This is fed by an ongoing cultural disconnect between emotional and physical pain. Western medicine tells us, for example, that *either* our pain is physical—in which case, we need to see a physician—*or* our pain is emotional, in which case we're sent to a psychotherapist. But neuroscience reveals that pain is never either/or; it is *always both*.

Psychological influences, frequently rejected and ignored, include some factors we've already seen in action—like perceptions, expectations, and the meaning assigned to sensations. This domain also includes some components we haven't yet explored, like thoughts, attention and distraction, and emotions. Emotional contributors to Sam's chronic pain, for example, included the depression and sadness that made his body hurt more, and the untreated stress and anxiety that kept his nervous system in overdrive. Emotions that affected Mateo's phantom pain included grief, shame, and hopelessness about his future. On the flip side, for both, hope and improved mood significantly contributed to pain relief.

Coping behaviors, or how we deal with pain once it starts, also live in the psychological domain of pain. These include the behaviors we decide to (or not to!) engage in when we wake up every morning—like whether we go for that walk, and how much alcohol and other substances we consume. Sam, for example, coped with pain by avoiding movement and activity, swapping soccer for videogames. This sedentariness, while normal and natural when we're in pain, made his body weaker, his joints stiffer, his muscles tighter, and his pain system more sensitive. A gradual return to movement and beloved hobbies was a significant part of Sam's healing protocol.

You'll notice that all of these "psych" factors overlap with "bio" factors, and this is critical. Because these divisions are merely lines drawn in the sand. When it comes to pain, everything is connected.

Pain Is Sociological

The third pillar of pain is the social, or sociological, domain of pain. I call this the "everything else" domain because of its immense breadth. This pillar includes all of the social, economic, and cultural factors that impact our health and well-being. This covers sex, gender, race, and ethnicity; socioeconomic status and access to care; social factors like isolation and social support; trauma and abuse, which play a more profound role in pain than we ever imagined; environmental stressors like poverty and racism; and larger contexts, including societal norms and expectations.

Social factors influencing Sam's health included social isolation, missing school, and falling behind socially and academically as a result. After being stuck at home for four long years, it's no coincidence that his first treatment goal was to reconnect with friends and rejoin his soccer team. The social domain of pain played a similarly significant role for Mateo. It was only when he finally had social support, coaching, and role models that he truly began to heal.

Together, all of these factors combine to create the pain we feel. No one ingredient is the sole contributor; all factors matter, all of the time. Pain is never—not ever—due to just one thing. This is extremely hopeful knowledge, because it means there are *many* things that can adjust pain volume beyond pills—and that we have much more control than we ever realized.

In fact, we get to put our hands directly on the dial.

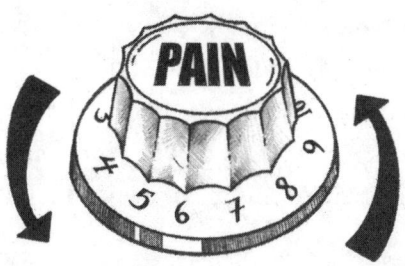

A Pain Dial in your central nervous system (brain + spinal cord) constantly adjusts pain volume.

Credit: Illustration by Eva Huzella

An Introduction to Your Pain Dial

Imagine that, in your central nervous system, you have a Pain Dial that operates much like the volume knob on your car stereo: Pain volume can be turned up, and pain volume can be turned down. This Dial serves as a control mechanism for our body's danger alarm. Over the course of an hour, a day, or a lifelong chronic illness, pain volume is always changing.

Many factors adjust pain volume. One of these, as we've all gratefully experienced, is pain medication. But there are many other factors that change pain volume, too, including our emotions.[vi,23] For example, when we're feeling stressed and anxious—our muscles are tense and tight, our thoughts are worried, we're predicting the worst—the amygdala and other parts of our nervous system send a message to the Pain Dial, amplifying pain volume. This is also true of other negative emotions, like sadness, hopelessness, anger, and frustration.

Cognitive factors like thoughts and attention are also powerful adjusters of pain volume. For example, when we focus on the body part that

---

[vi] In 1965, the founding fathers of modern pain science, Drs. Ron Melzack and Patrick Wall, developed the Gate Control Theory of Pain. This groundbreaking research incorporated emotional, cognitive, and sociological factors along with the neurobiological—forever revolutionizing pain science as we know it.

hurts, our brain's attention network sends a message to the Pain Dial amplifying pain volume. If you've ever noticed that pain seems to feel worse when you focus on it—when you stare at the needle going into your arm, anxiously awaiting pain onset, or when someone asks about your bad shoulder—you've experienced this phenomenon firsthand.

However, *the opposite is also true.*

When stress and anxiety are low—our muscles are relaxed, our thoughts are calm, and we believe our body is safe—the amygdala and other brain sites quiet the danger alarm. It isn't a coincidence that mindfulness and relaxation strategies can be powerful pain management tools (much as we might roll our eyes, myself included). This also helps explain the science of Lamaze breathing, shown to reduce the pain of contractions during childbirth.

Similarly, when our emotions are positive and our mood is high—we're feeling joyful and happy, or engaged in pleasurable activities with friends—the brain and nervous system turn pain volume down. As we'll soon see, emotions are biologically tethered to the brain's pain-control mechanisms, such that adjusting emotions necessarily adjusts pain, too.

Finally, when we're distracted—engaged in life and hobbies, or simply focusing on other things—the CNS lowers pain volume, quieting the danger alarm. Our pain is still there—it hasn't magically disappeared—but it feels less bad. If you've ever been so absorbed in some activity that you briefly forgot about your pain, that isn't magic: It's just your Pain Dial. This is one of the reasons why distracting a child with a screen can make an injection hurt less, and why athletes don't notice blood and bruises until the game is over.

Why is this so important?

Because pain takes away power. That's just what it does. Pain can make us feel like our bodies and lives are out of our control. Pain can steal our ability to work, exercise, play sports, have sex, enjoy hobbies, play with our grandchildren, spend time with friends, even walk outside in the sun. The Pain Dial reminds us that we have more control over pain than we ever realized—because there are many factors we can change to change pain.

Just as pain construction is multifactorial, pain *reduction* is, too. Having a sense of agency over our own bodies is critical to healing, and putting power back into the hands of people living with pain couldn't be more important.

Toward a Better Definition of Pain

Expanding our definition to include all the factors that comprise pain finally yields a clearer picture: Pain is our brain's *opinion* of how much danger we're in, based on all available data. Rather than an objective indicator of physical damage, we now know that pain is a predictive process constructed in relation to what's going on inside of us and around us in any given moment. This enables me to offer you an updated definition of pain: Pain is a biopsychosocial, subjective experience constructed by the brain, with inputs from our body and our environment, to make meaning of messages, predict what's to come, and determine how to best respond to ensure our survival.

Over the course of this book, you will learn all the details of this, and more: the science and complexity of this fascinating experience, and how, by unpacking pain's story, we can finally begin to heal.

3

If the Brain Can Change, Pain Can Change

What Goes Up Must Stay Up

Alex Honnold, world-class rock climber and daredevil, clung to a tiny crack in the granite. He was nearly three thousand feet in the air and still ascending, his toes pressed against the ledge. Far below, the trees looked like tiny, fuzzy stumps. He'd climbed sheer rock faces before, but this time, there was a catch. If he slipped, as he sometimes did, there would be nothing to stop his fall.

Because this time, he was climbing without a rope. And plummeting into the void below meant only one thing: certain death.

El Capitan, a famous monolith in Yosemite National Park, is beloved by nature lovers and rock climbers around the world. A massive vertical cliff, it typically takes experienced climbers between three to six days to complete. Using complicated rope systems, they spend multiple nights sleeping on portable ledges en route to the summit, suspended thousands of feet in the air. But Alex, ropeless, relying on his training, experience, and trusted climbing shoes, scampered up that impossible cliff in just under four hours—terrifying breathless onlookers, stunning the climbing world, and breaking the world record.

How did he do it?

Over the course of a year, Alex methodically trained on the face of El Capitan, safely secured to ropes to catch him should he fall. He attacked one small section at a time, committing every crag and crevice to memory. At the end of each day, he'd retreat to his van, parked at the base of the cliff—in which he ate, slept, and lived—to visualize and mentally rehearse the details of the day's training. He meticulously recorded every hand and foot placement in his notebook: Left foot goes here; right thumb goes there; big move on this section here. He even deliberately recalled the emotions he felt while on the rock—particularly the most dangerous, scary parts—to immerse himself in, and inure himself to, the fear. He dreamed of the route at night, until he could do it in his sleep.

By the time Alex was ready to free solo—to scale that vertical cliff without any protection or assistance—every single move was choreographed and memorized down to the most precise detail. Using the ink of rehearsal, Alex had tattooed the entire route onto his brain—making a mental map of that granite wall, synced with his movements. His brain and body were thus uniquely, exquisitely prepared for the task at hand.

Each of us has had some version of this experience, wherein practice and time seem to change the very fabric of our brains: how we perform, how we think, even how we perceive the world. This has major implications for pain. Because it is ultimately constructed by the brain, pain is never static or necessarily permanent. It can't be, because the brain is constantly changing.

And if the *brain* can change, *pain* can change.

Our Soft-Wired Brain: The Pain System Is Plastic

Regardless of what you may have been told—that chronic pain is "untreatable," that you are beyond hope, that you're doomed to hurt forever—neuroscience reveals that the brain is not hard-wired. The brain isn't fixed or permanent based on genes or biology. Rather, our brain and the rest of our nervous system—and, thus, our pain system—are "plastic," or changeable.[1,2] The term for this is "neuroplasticity": "plastic" meaning that the brain ("neuro") is flexible and malleable. Neuroplasticity is the ability of our neural architecture to adapt and change through learning, experience, and time.

In fact, our wonderfully plastic brain is *constantly* changing to suit our needs and our environment. Every time we learn something new—a new language, how to rock climb, Grandma's meatball recipe—our brain morphs and changes. And the more we practice any particular skill, the bigger and stronger the brain networks dedicated to that skill become. In neuroscience, this phenomenon is known by the principle "neurons that fire together, wire together."[i]

Similarly, every time we forget something—a colleague's name, the decimals in pi after 3.14, the year Columbus sailed the ocean blue—your brain also changes. Forgetting is the result of expendable neural pathways that are pruned just as we'd trim dead branches from a tree. In neuroscience, this phenomenon is called "synaptic pruning," otherwise known as "use it or lose it." Together, these concepts form the foundations of neuroplasticity.

Modifications in the brain occur on multiple levels. Microscopic changes occur in individual brain cells, called neurons, which have the ability to grow, shrink, move, and form entirely new networks depending on input and experience. Neuroplasticity can also produce large-scale changes to the overall map of our brain. The size and volume of particular brain structures like the hippocampus, a primary site of learning and memory, increase and decrease with time and experience, creating new pathways and shedding unused ones. Connections between brain cells, called synapses, also change, strengthening or weakening depending

---

i This is commonly known as the "Hebbian principle" after Donald Hebb, who helped pioneer a theory in the 1940s that explained neuroplasticity. However, the first use of this particular phrase was actually by two female neuroscientists: Dr. Siegrid Löwel, who wrote "neurons wire together if they fire together" in a 1992 article in the journal *Science*; and Dr. Carla Shatz, who summarized Hebb's work in a 1992 *Scientific American* article, writing "cells that fire together, wire together."

upon how frequently they're used. In this way, the brain is able to constantly adapt and adjust.

This flexibility is built into our very design. The myth that the brain consists of discrete structures, each assigned to just one function—for example, that our limbic structures *only* process emotions, or that the motor cortex only coordinates movement—is patently false. Rather, the brain is a dynamic, vastly interconnected organ that relies on these very interconnections, or neural networks, to do its job. And these networks are constantly being shaped and reshaped by new inputs.

The implications of this are astonishing. If we sustain a serious injury to our dominant hand in childhood and can no longer use it, the brain can rewire—and assign neural dominance to the other hand. If we lose our vision, the brain can reallocate networks that traditionally process vision to other senses, like sound, smell, and touch—making these senses stronger to compensate for the loss. And if a stroke damages brain circuits and sites responsible for movement, rendering us partially paralyzed, our brain can create *new* motor pathways—and we can learn to walk again.

Better yet, it isn't just our brain that's plastic: Our other biological systems adapt with time and experience, too. For example, our muscles grow and shrink with use, our hearts become stronger and more efficient the more we exercise, and record-breaking deep-sea divers can increase lung capacity until they can hold their breath for a full twenty minutes—while the rest of us landlubbers can only hold our breath for about thirty seconds.[ii] The fact that our brains and bodies are plastic means that *the pain system is plastic, too.*

In fact, this neuroplasticity underlies the brain's wonderful ability to change itself—something it does every single day of our lives.

---

ii Dr. Lorimer Moseley, pain researcher and scientist, refers to this as "bioplasticity." The many biological systems involved in pain production and reduction, including the CNS, immune, endocrine, autonomic, and other systems, all have evidence of plasticity. Source: Moseley, L. It's not just the brain that changes itself—time to embrace bioplasticity? IASP, n.d., https://www.iasp-pain.org/publications/relief-news/article/time-to-embrace-bioplasticity.

And the Cow Jumped over the Moon

Three years after her symptoms started, Fallon, fifty-four, wasn't any better. Instead of improving, she was getting worse.

Emaciated, pale, beautiful, and quiet, Fallon arrived at my office barefoot. Socks burned her skin, she said, and forcing her feet into shoes was unthinkable. She walked slowly, wincing with every step. The skin on her shins appeared thin and fragile, almost translucent, and was riddled with open sores. Despite frigid December temperatures, she wore shorts to prevent fabric from touching her sensitive legs, which were filthy. Nothing could touch them, she said—not even water. I later discovered she hadn't washed them for months, and had been bathing by submerging only her torso, hanging her legs over the lip of the tub to avoid touching the water.

The pain in Fallon's right leg had started with a seemingly innocent running injury three years prior. It radiated down her leg and back up again, her doctor ascribing it to possible nerve damage or entrapment. However, recently, and inexplicably, the pain had spread to her left leg. Both legs were now alternately red and blue, cold and hot, riddled with electric shocks of pain. Most alarmingly, her shins regularly erupted in weeping, bloody, eight-inch lesions. She'd been passed from doctor to doctor like a hot potato, and now found herself with a team of nine different physicians from nine different departments: primary care, dermatology, immunology, rheumatology, neurology, anesthesiology, cardiology, gastroenterology, and pain medicine. Despite the cacophony of opinions, she still had no clear diagnosis.

Finally, her ninth doctor had sent her to me.

Fallon sat gingerly on my couch, her hands folded in her lap. Her face was arranged into a pleasant smile, in stark contrast with what her body seemed to be expressing. Her legs were oddly still; over the course of an hour-long session, most people naturally bounce their feet, cross their legs, and fidget. But not Fallon. From the beginning of our appointment until the end, she didn't move her legs or even look at them. In fact, she seemed to ignore them entirely—as if they were so odious, so disloyal, that her brain was trying to forget them.

For the past year, Fallon had been undergoing multiday infusions requiring overnight hospital stays to treat the purported source of her pain, loosely described by her neurologist as an amorphous "autoimmune disorder." He managed this nameless condition with a variety of medications, including ketamine—a dissociative anesthetic with hallucinogenic properties, sometimes used as a horse tranquilizer. It was dissociative, her doctor explained, because it makes people feel detached from their bodies and surroundings. Sometimes, he administered morphine drips. Sometimes, Fallon said, she got both. The drug buffet temporarily eased her pain, but it inevitably came roaring back. The drugs also had some notable side effects that she dreaded. After the most recent ketamine treatment, she'd announced to her husband that she was a fish. He ran to call the doctor, who assured him that, despite Fallon's confusion, ketamine was a safe, commonly used pain treatment. But in spite of the various medications and infusions, nothing worked long term.

Fallon continued to decline.

In treatment, Fallon was guarded and aloof, sharing little beyond her hobbies. All I knew of her after multiple visits was that she was an artist, loved animals, and was a talented athlete who'd once been an Olympic-level archer. She'd retired her bow and arrow because of the pain, she said, but missed it terribly. Beyond that, I knew nothing. The lack of alliance also meant that I wasn't making much headway—I was failing as a pain detective. So, in an attempt to know her better, I asked if she'd be willing to show me her art, hoping it might offer some clues. The first painting she shared instantly sent chills up my spine.

The painting featured an inky night sky with a pale spot where the moon should've been. A cow was attempting to jump over a yawning canyon, from one precipice to another. It was pictured mid-leap, suspended over a dark abyss, moonlight framing her silhouette as she hung in the air. A line from a children's fairytale danced through my head like a song: "And the cow jumped over the moon, and the dish ran away with the spoon." But something was terribly wrong.

The cow appeared to be bleeding profusely from her chest. Thick, red

Credit: Anonymous, used with permission

blood poured into the canyon below. In the bottom corner, nearly invisible, a shadowy, faceless figure with a black beret lifted a loaded gun to the sky. The cow was desperately trying to complete her journey, to get from one side to the other…but she wasn't going to make it.

This was no children's story. Despite Fallon's calm, impassive face, she was desperately communicating that she was not okay.

When I shared this observation, Fallon's eyes started to tear. It was so subtle I almost missed it. But it was a breakthrough, and over the next few weeks, she finally began to share. I learned that many factors, previously undisclosed to anyone, were contributing to her pain—hijacking her pain system and launching her body into emergency mode. For one, she'd recently been fired from her job as a result of repeated, extended medical leaves. It couldn't have come at a worse time: Her expensive medical care had drained the family bank account, and money was extremely tight. Her home was now under foreclosure. She felt like a burden. She'd stopped being the wife and mother she aspired to be—she couldn't take her son to school or cheer from the soccer sidelines, and sex with her husband had been out of the question for nearly a year. She rarely left the house these days, or even bothered to change out of pajamas. Her life had lost its color.

Only one chronic pain syndrome fit Fallon's symptoms, a rare condition her entire team had missed: complex regional pain syndrome, or CRPS, also known as the "suicide disease." It made sense that it wasn't on her team's radar. It also became clear that her team had been interpreting her symptoms according to their siloed specialties, focusing on specific systems and body parts. The dermatologist, for example, believed it was a skin issue; the neurologist, a nerve issue; the immunologist, an autoimmune issue. But CRPS is a *whole-person* issue. CRPS typically affects limbs and extremities—an arm, leg, hand, or foot—and is characterized by certain telltale signs: changes in skin temperature, as the limb can become abnormally cold or burning hot; changes in skin color, most frequently blue, purple, red, or white; and changes in skin thickness, such as becoming thicker or, conversely, fragile, dry, and prone to tears. CRPS also has the rather unique and infuriating characteristic of occasionally "jumping" from one side of the body to the other.

Fallon presented with some other common symptoms of CRPS, too: swelling, muscle weakness, peripheral neuropathy (an electric, sharp, radiating pain in her legs and feet), and alarming changes in skin sensitivity. She'd become hypersensitive to touch and other nonpainful stimuli, a condition known as allodynia. This explained why she couldn't wear socks or even bathe her legs. She also experienced hyperalgesia, or an abnormally heightened sensitivity to painful stimuli, like a needle prick. But the symptom that fascinated me most was her "selective neglect": a tendency of some CRPS patients to unconsciously ignore the affected body part, resulting in disuse and disengagement with the affected limb. This was why Fallon failed to move her legs and feet in session, even over the course of many hours. It wasn't on purpose; she didn't even know she was doing it. Patients sometimes report feeling like the affected limb isn't even part of their body. And in extreme cases, patients tragically decide to have the limb surgically removed.

Despite its scary name, the fact that the syndrome was known—as was its treatment—cheered Fallon immeasurably. She was no longer a medical mystery, and being "seen" felt medicinal. Moreover, the news was revelatory: while hers was indeed a leg issue, it was also a *brain* issue. She quickly absorbed all the science with great interest. CRPS, she learned, is believed to develop when the brain and nervous system become hypersensitive to sensory messages from the body, and start misinterpreting nondangerous input as dangerous.

These amplified messages in the peripheral and central nervous systems generate a massive alarm response, accompanied by changes in the immune, circulatory, musculoskeletal, endocrine, and other bodily systems. Her pain flares, she learned, were triggered and maintained by a variety of factors—not just her old running injury, but also environmental stressors, inactivity, isolation, depression, and trauma. Knowing the science helped her see a path forward.

Fallon's physical therapist and I joined forces to help Fallon identify all the factors triggering and maintaining her pain, and to brainstorm how to make changes. To increase connectivity between her brain and her neglected, painful legs, and to help her desensitize to sensory input, I asked Fallon to gently touch her legs and tap her feet every few minutes in session. To reduce her fear of movement, Fallon engaged in daily PT exercises to gradually increase strength, balance, and range of motion. Every day, Fallon went for short walks outside in the sun, and dutifully applied cream to her sores. She progressively exposed her legs—and subsequently, her brain—to a variety of sensations using different textures and temperatures for few seconds at a time: a soft cotton towel, warm water, and, later, a transcutaneous electrical nerve stimulation (TENS) machine: a small device that delivered a gentle tingling sensation that she could make gentler or stronger as she wished.

Inside Fallon's malleable brain, her pain system was calming and desensitizing—slowly learning that there was no emergency, that touch and movement weren't dangerous, that the amplified alarm wasn't needed.

After one month, Fallon was able to tolerate light touch, and then longer periods of touch. With gradual exposure to water, she was able to submerge her legs in the bath for a few seconds—then a full minute, then fifteen minutes. After two months, Fallon regained some mobility, strength, and confidence in her legs. The rest of her body was responding, too: inflammation started calming, her circulation improved, and the lesions healed. Her skin started looking healthier and less translucent.

After twelve weeks, her legs stopped freezing and burning, stopped turning colors, and started looking healthy again. All the while, she painted, went for walks with her husband, and invited friends over for movie nights. The support was medicinal. As she used the pain-control strategies she learned in treatment, she began to feel better—not just physically, but also emotionally. Hope returned like a full moon on a dark night. After twelve weeks of treatment, Fallon still had occasional flares, but felt well enough to resume work part-time. Her depression started to abate, and her anxiety calm. Sleep improved. As pain volume reduced, Fallon started reporting less pain, and then days with no pain. She went back to archery—at first, just to visit friends and watch them practice; then, to hold a bow; and finally, when she felt strong enough, to participate in weekend practices.

And then, just as unexpectedly as she'd arrived, Fallon suddenly didn't need me anymore.

How Pain Becomes Chronic

Fallon's story is a stark example of the ways in which the brain can wreak havoc on the body, and helps us understand how pain can become chronic. Indeed, that question is one I'm often asked: "If tissues heal within three to six months, why am I still in pain?"

It's a good question and a fair point: Sometimes pain persists long after injuries have healed. It's a good reminder of pain's ultimate location; after all, we perceive with our brains, not with our skin.

While there are multiple ways by which pain can become chronic, among the most studied and well-known mechanisms is central sensitization—"central" due to the role of the central nervous system (CNS), and "sensitization," which describes an increase in CNS sensitivity and reactivity with experience and time. Evidence of this process at work has been found across a wide swath of chronic pain conditions, including Fallon's illness, CRPS, as well as cancer, back pain, headache, arthritis, sickle cell disease, neuropathic pain, postsurgical pain, autoimmune disorders, hypermobility syndromes, pelvic pain, dental pain, orofacial pain, fibromyalgia, and irritable bowel syndrome.[3,4] In fact, central sensitization is believed to play a role in the chronicity of *all* pain to some degree.[iii,5,6]

While complex, this phenomenon has its roots in neuroplasticity. The pathways in our brains are like the muscles in our body: The more we use them, the bigger and stronger they get. For example, the more Alex practiced climbing El Capitan, the more he used the "climbing pathway" in his brain. And the more he used it, the bigger and stronger it got, until he'd created a dynamic super-map of that cliff. For me, it was the piano. The more I practiced Chopin waltzes and Pachelbel's Canon, the bigger and stronger my "piano pathway" got, until I didn't even need to look at the sheet music—my fingers just knew what to do.

Just as the brain can become good at mountain maps and piano playing, it can also get really good at pain.[7,8] When we have pain for weeks, months, and years, our brain and spinal cord inadvertently "practice" pain. The longer our CNS practices pain, the bigger and stronger our pain pathways become.[iv] The stronger these pathways get, the more *sensitive* our pain system becomes.

---

iii "In fact, so much has been learned about the ubiquity of central sensitization (CS) in populations of chronic pain patients that there is now widespread consensus in the pain field that CS might represent a...core mechanism for pain, above and beyond nociceptive and neuropathic pain mechanisms that have been recognized for some time." Source: Harte, S., Harris, R., & Clauw, D. (2018). The neurobiology of central sensitization. *Journal of Applied Biobehavioral Research, 23*(2), e12137.

iv Due to the complexity of pain science, I simplify concepts by referring to our "pain system" and "pain pathway"—which are in fact a collection of brain networks and body systems that work together to construct the pain experience. There is no one single, universal "pain pathway" in the human body.

What does this mean? A highly sensitive instrument is one that picks up small inputs missed by less perceptive machines. For example, a sensitive earthquake detector, called a seismograph, is considered superior because it can detect faint rumblings beneath the earth far earlier and more accurately than less-sensitive machines. Some animals, like foxes and dogs, have highly sensitive hearing: Their ears pick up faint sounds humans can't detect, magnifying auditory input and perceiving it more strongly. This means that sounds that seem loud to us are deafening to dogs—which is why they hide under beds across America during Fourth of July fireworks.

Similarly, our brain can become a highly sensitive instrument—in ways that help us and, in the case of chronic pain, in ways that don't. Scientists have found that pain has the ability to alter our nervous system, turning the brain and spinal cord into sensory super-detectors.[9,10,11] The cells and pathways in our pain-transmission system, research shows, get better and better at making it over time. This sensitization can occur at every level of the pain system, starting with the signaling and sensitivity of our nociceptors, "potential danger" detectors in our tissues, to the cells and pathways in our spinal cord, all the way up to the brain—a perfect example of full-body plasticity. When the nociceptors in our tissues, bones, joints, and organs become sensitive and hyperresponsive, this is known as "peripheral sensitization." All told, a sensitive pain-transmission system is more responsive and reactive, has a lower pain threshold, and requires less input to trigger the danger alarm.

In short: the longer we have pain, the better we get at producing it.

Warning messages are now interpreted as signs of danger *even when our bodies are actually safe*. As such, small bits of sensory input from our bodies are amplified, resulting in more pain with less provocation.

Lest we get angry at our bodies for doing such a thing, sensitization isn't actually a bad thing on the whole. This learning process is simply our body's way of becoming more efficient at responding to potentially dangerous stimuli—and adapting to our ever-changing environment, as the body is designed to do. For example, vaccines teach our immune system to become more "sensitive," or more efficient and reactive, to dangerous pathogens, so that future invasions can be battled more quickly and successfully. So it is with pain: Past experiences train our CNS to respond more quickly and urgently to similar, potentially dangerous inputs.[12] And while this can be a lifesaving adaptation in the short term, it isn't particularly useful over the long term. In fact, over time, it can even work against us.

When pain becomes chronic, the pain-transmission system gets stuck in a sensitive state rather than snapping back to baseline.[13] This prompts our pain system to start generating pain in the *complete absence* of danger messages from our tissues.[14,15] These warning signals are therefore no longer useful, nor do they protect us. Consider, for example, the pain Fallon felt when she tried to bathe: Water isn't dangerous to someone with CRPS, but her pain alarm shrieked anyway. Chronic pain is thus considered maladaptive—reflective not of tissue damage, but of an alarm system gone awry.[16,17,18] This is true whether we experience chronic pain in our head, shoulders, knees, or toes.

Of note, sensitization isn't the only process at play here. When pain becomes chronic, systems across the brain and body are affected: from the tiny cells of our immune system, to our hormone system, to the very size and functioning of our brain structures.[19]

The Highly Sensitive Person

Given the importance of sensitivity to pain processing, it's important to note that sensitivity, while influenced by environment and experience, is also, at least in part, genetic. While there is no one single "sensitivity gene," this trait appears to be heritable, passed down from parent to child, with a range of sensitivity levels across the human population. Data suggests that about 20 percent of us are highly sensitive, with brains that process physical, emotional, social, and environmental data differently from less-sensitive individuals. Highly sensitive people (HSP) can experience all types of sensory input differently, including sound, taste, and touch, and can often sense, and sometimes even feel, the emotions of everyone in a room. This has major implications for pain: HSP reportedly have lower pain thresholds, along with a reduced pain tolerance.[20,21] This suggests that, at baseline, HSP may have a particularly sensitive and reactive pain system. Moreover, people with this sensitivity trait, under certain circumstances, may be more likely to develop central sensitization—and chronic pain.[22]

The Power of Neuroplasticity

But here's a critical piece of good news: neuroplasticity got us into this mess, and neuroplasticity can get us out of it. Just as the pain system can become more sensitive and responsive, it can also become *less*. In fact, because our brains and bodies are plastic, *we can actually retrain our pain system*. It's possible to create new brain pathways, and a whole lot of them: pathways that lower the pain alarm, pathways that change the meaning of the message, pathways that reduce our sensitivity, and even pathways that increase our functionality, enabling us to resume life and the activities we love.

But how?

A great way to kick-start the process, as Fallon and Sam experienced firsthand, is by helping the pain system desensitize. One way to do this is via gradual exposure to small doses of potentially triggering stimuli such as movement, touch, and activity.[23] Over time, desensitization can decrease our sensitivity to pain, increase our pain tolerance, and reduce pain-related fear. While this process may sound lengthy and complicated, in reality, our brains sensitize and desensitize to stimuli all the time. For example, while sitting in a dark movie theater, our eyes and brain sensitize to light as they adjust to the dark. This enables them to pick up and maximize every small bit of light. When the movie is over and it's time to return to the sun, the best (and least painful) way to readjust to bright light is to gradually open the blinds, letting in just a little bit of light at a time. Slowly, slowly, our eyes and brain adapt and desensitize, until presto—we are again in a light-filled room. This is the beauty of plasticity, and the science behind Fallon's recovery.

Unfortunately, the opposite is also true: not moving, staying inside, and overprotecting the painful body part keep the brain and body in a state of "darkness"—confusion, fear, and hypersensitivity. While helpful for acute pain, prolonged periods of inactivity and withdrawal make chronic pain worse, because our nervous system stays sensitized to, and focused on, the pain. The false-alarm system keeps shrieking loudly without anything to control its volume. The longer we avoid activity, the longer our brain remains in this protective state—and the longer pain will last.

Ultimately, there are many ways to retrain the brain and change pain.[24] Critical steps include:

1. Rewiring the nervous system by changing the way we move, eat, sleep, think, and socialize.
2. Changing the meaning of the message by reframing chronic pain as an amplified alarm unlikely to indicate danger.
3. Gradually exposing brain and body to small doses of sensory input, encouraging the creation of new brain pathways linking activity and movement with *safety* instead of danger.

This creates an "off button" for the brain's pain alarm that, when pushed repeatedly, weakens the association between "chronic pain" and "danger"—pruning that faulty neural pathway, and turning the volume down on pain. You will learn a zillion strategies for how to do this in Part III: The Pain Protocol.

Stories of hope and recovery abound. And not just from the pain of rare diseases like Fallon's, or chronic illnesses like Sam's—but also relief from the pain of injuries and accidents, bad backs and broken bones. One such story belongs to Todd.

When his truck was totaled in a head-on collision, Todd, fifty-seven, sustained a litany of serious injuries—among them two neck fractures, a broken leg, shattered heel bones, and a traumatic brain injury that caused a brain bleed.[25] Prior to the accident, Todd had been a marathon runner.

Afterward, he was forcibly bedridden—debilitated with crippling pain, told he'd likely never run again. But Todd and his wife decided they didn't like that prognosis. So, after taking the better part of a year to heal, Todd slowly started training again, taking on one small goal at a time. At first, the pain was so unbearable that he could move for only a few minutes before having to stop. But every day he worked at it, reshaping and strengthening his brain and body, until he could walk again—then jog, and then run. He focused on his diet, strengthening exercises, and healthy daily routines, and bolstered his support network by hiring a running coach.

Recently, Todd was finally able to start running long distances again. Inspired by his progress, Todd and his wife registered for their next race. Only this time, it wasn't a marathon.

In just under nine days—that's days, not hours—Todd and his wife completed an *ultramarathon*, a full 314 miles, on foot. And if you think that's miraculous, consider this:

Todd no longer feels pain when he runs.

4

The Pain Recipe

If we want to understand why we have pain when we have it, let alone exert some control over it, we must first identify the ingredients that go into creating it.

Just as there's a recipe for brownies—particular ingredients, added in a particular order, baked at a particular temperature, for a particular amount of time—there is also a recipe for pain.

You may already know this intuitively. Perhaps you already know what triggers a bad pain day. My body, for example, inevitably feels worse when I sit without moving for long periods of time, sleep poorly, eat crappy foods, and am overwhelmed by stressful deadlines and responsibilities. The contributing culprits for Sam's crippling migraines included poor nutrition, school stress, eyestrain, and depressed mood. For Mateo, isolation, hopelessness, and a confused brain map kept pain dialed high. Fallon's pain triggers and amplifiers included stressors like unemployment and finances, immobilization, and a pain system stuck in emergency mode. This is what I call our "high-pain recipe," and everybody has one. Each person's recipe is different, but when combined, this unique mix of ingredients, or triggers, combust to create, perpetuate, and exacerbate pain.

To construct a high-pain recipe, we need ask only a few basic questions: When we have a bad pain day and our bodies feel awful, what makes it bad? What factors seem to trigger flares and amplify pain volume? To solve this puzzle, we lay out all the possible ingredients in our Pain Menu (see table). This biopsychosocial menu consists of anything that has ever contributed to your pain, and everyone's is different: tissue damage or disease; genetics, like family history of migraine; physical activity (or lack thereof); sleep issues; diet; trauma or abuse; emotions and mental health; unhealthy or toxic relationships; and environmental stressors like work, finances, racism, or a global pandemic.

PAIN MENU

Bio	Psych	Sociological
Genetics, hormones, brain chemicals	Emotions	Culture
Age and sex	Coping strategies	Gender
Tissue damage	Predictions and expectations	Access to care
Inflammation	Thoughts	Race and ethnicity
Muscle tension	Meaning assigned to sensations	Socioeconomic status
Diet, nutrition	Mental health	Environmental stressors
Exercise	Memories	Social support and relationships
Sleep	Beliefs	Trauma

The ultimate redeeming quality of our high-pain recipe is that it naturally leads us to its opposite: our *low-pain recipe*. These are the ingredients that go into creating a day with less pain. These recipes offer a road map for healing—because, as with any good recipe, the ingredients in our pain recipes can be tweaked and adjusted. While we cannot change some ingredients, like the harsh reality of aging or the environment in which we were raised, there are dozens we can exert control over: how much we move, what we eat, our sleep habits, and even our expectations, thoughts, and mood. All of these inevitably affect how our bodies feel.

Each high-pain ingredient thus becomes an opportunity for intervention. If stress contributes to flares, we can build stress-reduction strategies into our treatment plan. Worsening pain after a night of poor sleep has an antidote: a strategy called sleep hygiene, which you'll learn about in Part III. If isolation hurts, we can find ways to create more social connection. Inflammation and swelling may require ice, rest, and medication. In fact, the more high-pain ingredients we replace with low-pain ingredients, the better we're likely to feel.

Understanding that pain is biopsychosocial gives us incredible agency over our bodies—the ability to make new decisions, take action, and change the pain we feel. True pain medicine, it turns out, is *many* things, and we've carried these solutions inside us all along.

Welcome to the Pain Recipe Method of pain management.

If There's a Recipe for Brownies, There's a Recipe for Pain

Joyce, age sixty-eight, mother of five and grandmother of nine, was an avid consumer of memoirs, a compulsive collector of yellow watering cans, and a former ballroom dancer. She was a chef like her mother before her, and ran her own catering business. Money was tight, but she was very proud of the life she'd built.

Her chronic back pain started after what her kids referred to as "the salsa incident" twenty years ago. She'd been salsa dancing all afternoon with her impossibly handsome now-ex-husband, also a dancer. Things had been tense between them; he'd been distant and disconnected, and Joyce suspected he was having an affair. She'd been exhausted that day, she remembered, both physically and emotionally; her body was fatigued, and she felt drained. But she wanted to keep up, to not seem as old or worn-out as she suddenly felt. In one moment—one, single moment she'd replayed countless times since—her back inexplicably spasmed and gave out. Skirts swirling, bracelets flashing, she crumpled to the floor, unable to get back up. Her back hadn't been the same since.

Joyce's orthopedist had diagnosed her with degenerative disc disease. On the MRI scan, he'd circled the evidence of her compressed, herniated discs in red marker. Her discs had degenerated over time, he said, and her vertebrae were now so close together that they were rubbing and "eroding" one another. Joyce's doctor had further informed her that she "had the spine of an eighty-year-old," and said that if she didn't have surgery, she'd end up paralyzed. She was only forty-eight at the time.

Without hesitating, Joyce had surgery. Her then-husband had grudgingly agreed to take care of things while she recovered. But on day two, she groggily awoke to find him scurrying out of the bedroom with a suitcase and an armful of photographs that fluttered to the floor behind him like dead leaves. He hadn't even had the cojones to tell her he was leaving. *Great dancer*, she noted; *complete coward*.

The split was confusing to her kids, and ultimately changed family dynamics. While she still saw her daughters, her son grew distant, and then stopped communicating altogether. His inexplicable estrangement broke her heart—even more so because there was nothing she could do about it. The compound grief of losing her son after losing her husband sometimes felt like more than she could bear.

In the midst of all of this, her body struggled to heal. Her doctors said everything "looked fine" structurally and that the pain should diminish within a few weeks of surgery. But it only got progressively worse. Since the surgery, Joyce had struggled to dance, walk, work, even stand. Her flares seemed random, coming and going without warning, reducing her to weeks in bed. So, she did what she thought she needed to do: She stopped dancing. She went out less, and stayed off her feet and out of the kitchen. Seasons passed. Her body became heavier and more cumbersome. She could barely keep her business afloat, and worried about it going under.

Her life got smaller. Her world contracted.

In the two decades since, she'd had more surgeries, invested in years of physical therapy, and sat for regular steroid injections and chiropractic adjustments. A spinal cord stimulator had been implanted in her back. She'd been prescribed countless pain medications, and had tried various diets to lessen the burden of her suddenly heavy body on her suddenly lame frame. On top of everything, Joyce was aging, and her body was showing the signs. Joyce gathered reams of advice from professionals, friends, and the internet, all of which she dutifully recorded by hand in a yellow, spiral-bound notebook. She saw providers she loved, and just as many she could do without. But despite years of interventions, her pain persisted.

It was frustrating—she felt like she'd tried everything. Finally, when they could do no more, her doctors gave her a new diagnosis: failed back surgery syndrome. She was bewildered; hadn't the surgery failed her, and not the other way around? Nevertheless, her doctors suggested she stop trying to fix it, and get a wheelchair. That's when Joyce found me.

When she learned about the pain recipe, the concept changed her life. A lifelong cook, Joyce appreciated the metaphor: No single ingredient is ever the most important in any recipe. There's no star, no MVP. As every chef knows, all ingredients work together to create the magical alchemy that produces the desired end product. A great fudge brownie, announced Joyce in session, doesn't rely upon good chocolate alone; if you leave out the flour, sugar, and eggs, all you'll have is cocoa powder. Leave out the cocoa, she scoffed, and you lose the "brown" in "brownie." Environment, tools, and temperature matter, too: Brownies baked in too large a pan emerge hard and bricklike, while brownies baked at too cool a temperature are raw. The key elements in a recipe aren't just the groceries—but the context and the environment, too.

Joyce spent the next few months tracking all of her pain ingredients across physical, emotional, social, and environmental domains of life. There was no one thing that predicted her pain, she discovered. This was clearly evident once she started tracking it: There were always fluctuations, even within the same day, even though her degenerated discs remained the same. Therefore, she reasoned, something else must be adjusting the dial. It wasn't just her old injury creating the pain; not just the process of aging, her insomnia, nor her propensity for sitting on the couch for six-hour stretches watching *Law and Order*. *All* factors mattered: the things going on inside of her, combined with the things going on around her. Her loneliness, her estrangement from her beloved son, the horrific events on the news, mounting unpaid bills—a pile of stressors that created a dull hum in the background, over which she had little control—those were important, too.

Her high-pain recipe quickly revealed itself once she knew what to look for; and her low-pain recipe, it turned out, was just the opposite. As soon as Joyce was able to identify the ingredients, she was able to start changing her pain. It was no quick fix. But as with every other friendly recipe she'd ever encountered—her mother's banana walnut muffin recipe, her cheddar quiche recipe—the recipe itself was a road map and a guide, telling her exactly what she needed to do in order to achieve the results she desired.

In session, Joyce was dedicated and motivated. The longer she worked at it, the more she found new, creative ways to lower pain volume by tweaking different ingredients. To improve her sleep, I offered her a sleep hygiene protocol, which she dutifully implemented every night. After reviewing her diet together, Joyce committed to revamping it, and started eating three healthy meals each day rather than snacking on fast food and chips. We established a pacing plan to help her gradually strengthen muscles and regain mobility: Every hour on the hour, she went for a five-minute walk in the sun. She began delegating work responsibilities to her daughters to reduce stress and overexertion. She stopped watching the news, and joined water aerobics at the community pool instead. Dancing in the water was a revelation and a joy: It was low-impact, she felt little pain, and she even made some friends.

As part of her commitment to prioritizing emotional health, Joyce invested three months in therapy—the max her infuriating insurance company would cover. She never imagined that therapy could help her pain. In fact, she'd avoided it for years, thinking it silly and irrelevant given that she wasn't "mentally ill." But its impact, once she started, was undeniable. Joyce even tried mindfulness, which helped her tune in to her body and recognize when she was clenching and bracing her neck, shoulder, and back muscles—habits that only made her pain worse. For her, processing emotions while learning skills to manage stress, grief, and loss were among the missing, magic ingredients.

Each week, Joyce wrote three recipes on her kitchen whiteboard as a reminder and guide:

1. The recipe for the dessert she intended to bake that weekend for her grandkids.
2. Her high-pain recipe.
3. Her low-pain recipe.

As Joyce wrote her weekly recipes, she visualized the steps she'd need to take to ensure her day was full of as many low-pain, and as few high-pain, ingredients as possible. The more mindful she was about these daily ingredients, the more successful she seemed to be. This morning's whiteboard read:

JOYCE'S PAIN RECIPE

Fudge Brownie Recipe	High Pain Recipe	Low Pain Recipe
1.5 cups flour	Inactivity/overexertion	Hourly stretches, water aerobics, daily short walks with rests
2 cups sugar	Taking on too many responsibilities	Setting boundaries, delegating work
4 eggs	Skipping meals, processed foods	Three healthy meals a day
1 cup butter	Poor sleep	Sleep hygiene
½ teaspoon salt	Life stressors	Relaxation strategies, consuming less news and social media, protecting self-care time

¾ teaspoon baking soda	Grief + sadness over her divorce and estrangement from her son	Therapy, journaling
½ cup cocoa powder	Loneliness	Social plans twice a week
1 tbsp vanilla	Belief that pain always equals damage	New belief: Chronic pain doesn't necessarily mean damage; it can indicate a body out of balance and an overprotective pain system on high alert

The more Joyce followed her recipe, the more functional she was, and the lower her pain volume. Motion provided lotion to her painful, swollen joints. Her muscles became increasingly stronger and more limber, and circulation improved. Better nutrients and sleep provided better building blocks for healing. As she made more social connections and spent more time outside, brain levels of feel-good neurotransmitters like dopamine, serotonin, oxytocin, and homemade opioids rose. Her body churned out fewer stress hormones, and inflammation calmed. Slowly, her pain system desensitized. She still had bad days, sometimes inexplicably. On those days, the usual strategies didn't work, and she needed to once again tweak her recipe or simply rest. But despite all obstacles and setbacks, Joyce had her life back, and she felt alive again.

The biggest perk, aside from being able to dance again, was getting back to baking in her warm, fragrant kitchen with her precious grandkids. After twenty years of debilitating back pain, Joyce finally experienced some relief.

Your Pain Recipe: How to Use This Book

This concept offers a completely new way of looking at pain. No longer is it solely the product of injury or illness. Now, pain is appropriately complex—and also simpler and less mysterious, because the ingredients are discoverable. And I am going to help you find them.

The Pain Recipe Method of pain management helps us uncover four things critical to pain reduction:

1. Pain triggers and amplifiers
2. High-pain ingredients that are in our power to change
3. Opportunities for intervention
4. A road map for creating a new, low-pain recipe

In this book, you will learn how to identify the ingredients in *your* unique pain recipe, along with strategies for how to change them. You will learn more about how the science of these biological, emotional, cognitive, social, and environmental factors drastically affect our brains and bodies, and how to use this knowledge to your benefit. As you read Part II, be on the lookout for all of the ingredients that might apply to you. Some may initially be invisible, like the baking soda in the cookie—tasteless, yet critical. Trauma is often one of these factors, as we will soon see. But as you use this book as a resource and a guide, your recipe will emerge, as well as guidelines for effective treatment.

Because it is only by knowing our pain recipes that we can harness neuroplasticity, change the ingredients, and finally begin to heal.

PART II

The Pillars of Pain

The most critical tool for treating pain, it turns out, is understanding it. Without this broader lens, every time we hurt, we go looking for damage—and our quest for healing rarely goes beyond the doctor's office. But a new understanding of pain's building blocks *finally* inspires new interventions. A biopsychosocial approach, while perhaps unfamiliar, is incredibly optimistic. Because it means that we have more ways of treating pain than we ever imagined—and these untapped avenues offer an incredible amount of hope.

In Part I, I introduced the three pillars of pain: the biological, psychological, and sociological ingredients that interact to increase—and decrease—the pain we feel. We spent time exploring the first pillar, the biological domain, including the basic neurobiology of how pain works. These physiological, anatomical, and chemical ingredients, from tissue damage to cellular mechanisms, are fundamental. They are also the most familiar and frequently treated.

In Part II we'll dive deeper, breaking down the two remaining pillars of pain—psychological and sociological—into their component parts. The psychological domain is divided into emotional and cognitive ingredients in Chapters 5 and 6. We'll learn the many ways that emotions, like joy and sadness, and cognitive factors, like thoughts and attention, adjust our Pain Dial. I answer the questions I get asked most, such as: Can heartbreak hurt as much as a kidney stone? Why do we get "butterflies" in our stomach when we're nervous? Why does fearfully anticipating pain seem to make it worse, while distraction makes things hurt less? And is pain ever all in our heads?

Chapters 7 and 8 tackle the sociological pillar of pain, divided into social and environmental ingredients. We'll address the fascinating science behind why we instinctively rub our knee after bashing it, how human touch quiets the pain system, and how social isolation can make us sick. We'll learn the surprising ways that environmental factors like trauma amplify the pain alarm, why Greeks and Romans worshipped pain gods, and how our sex affects the pain experience. Chapter 9, "The Body's Pharmacy," ties all this science together, exemplifying how these three pillars constantly interact to transform the pain we feel. As I unpack the science, you'll see for yourself what's really happening when the body makes pain—and also the processes that help us heal. Patient stories help guide the way.

You'll learn how all of these ingredients—from how much we sleep, to the emotions we feel, to the people we see—constantly combine in different arrangements to alter the pain experience. Indeed, these varying combinations are the very reason pain fluctuates over the course of a week, a day, even an hour, and why ten people with the exact same injury report ten completely unique pain experiences. Of course, this is a paradox: While the biological, emotional, cognitive, social, and environmental factors you're about to discover are inextricably interconnected, we must separate them out and break them down to really see how they work.

By the end of Part II, you'll better understand the hidden ingredients in your *own* pain recipe. And this will give you something you've never had before:

a road map for healing.

5

Pain Is Emotional

RECONNECTING BRAIN AND BODY

The Biology of Broken Hearts

James, a creative six-year-old, loved comic books, storytelling, and playing pirates with friends. Not long after his sixth birthday, his older brother—his mother's favorite—was killed in a horrific accident. His mother was grief-stricken. She fell ill and became bedridden. One morning, when James went into her darkened room seeking comfort, she mistook him for his lost brother, calling him by the wrong name. In the hopes of healing her, James began wearing his dead brother's clothes, and even adopted his whistle. But despite his best efforts, he couldn't cure her. He later wrote that his mother took comfort in knowing that her lost son would remain a boy forever, and never grow up. Stricken by the stress and sadness of his brother's death, his mother's ailing health, and his own grief, James suffered serious physiological effects. Despite having no illness or disease, James's growth mysteriously stopped—a condition known as psychogenic dwarfism, or stress dwarfism. Scientists concluded that extreme stress and emotional deprivation had hijacked James's biology, slowing the production of growth hormones and stunting his development.

This may sound fictional, but it is all too real. You've likely heard of this man. He is none other than J. M. Barrie—the acclaimed author of *Peter*

Pan, the timeless childhood tale about a boy who lives in Neverland, fights pirates, and never, ever grows up.

While stress dwarfism is rare, the impact of emotions on human health is not. Our risk of having a heart attack, for example, increases twenty-one-fold the day after a loved one dies. This condition, known as "broken heart syndrome" or "stress-induced cardiomyopathy," is triggered by sudden heartbreak and grief, and it can happen to anyone—even those with otherwise healthy hearts. Other emotions, like agitation and excitement, also affect our hearts and health: On days our favorite teams compete in sporting events, we're up to three times more likely to suffer cardiac emergencies, including heart attacks and death, compared to days they don't play.

It may seem astonishing that emotions can affect our bodies this way. But this chapter will demonstrate why emotional health is central to the science of pain. In fact, if we aren't taking care of our emotions, we aren't taking care of pain.

Emotions Are Physical

The question I get asked most often is "Do you treat physical pain, or emotional pain?" My answer, without hesitation, is always "Yes." This is because of one simple fact: Emotions aren't just in our heads. They also live in our *bodies*. You've experienced this yourself a million times: Nervousness before a talk or test might come out as sweaty palms, a pit in your stomach, or nausea. While watching a scary movie, fear might manifest as a racing heart or goose bumps. When we're sad, salt water leaks from our eyes. Depression slows us down, saps our energy, and makes our bodies feel heavy. Stress can cause breakouts, hives, chest pain, palpitations, and can make our hair fall out. Major stressors have even been known to alter menstrual cycles, induce premature childbirth, and trigger miscarriages.

Emotions constantly and profoundly impact our bodies, whether we realize it or not. In medicine, this is called the "somatic," or bodily, aspect of emotions, and no human body is exempt. The list of how emotions manifest physically is endless. Fascinated by this seemingly forgotten fact, I've spent the past decade interviewing patients about the various ways in which their emotions manifest physically. Together, we've generated a list of over eighty options. As you consider some of theirs, consider yours:

EMOTIONS IN THE BODY

- Nail biting, finger picking
- Lightheadedness, dizziness
- Dry mouth
- Fatigue, exhaustion
- Stomach "butterflies"
- Tight jaw, grinding teeth, TMJ
- Stuttering, difficulty finding words
- Change in appetite, eating more or less
- Bloody nose
- Irregular menstrual cycle, missed periods
- Urge to use the bathroom
- Restlessness, fidgeting: bouncing legs, tapping feet
- Headaches
- Sweating
- Rapid heartbeat, heart palpitations, heartburn
- Stomachaches, nausea, vomiting, diarrhea, constipation
- Rashes, hives, breakouts
- Talking faster or slower
- Cold sores (herpes outbreaks)
- Body pain
- Hair falling out or turning white
- Erectile dysfunction, impotence, low sex drive
- Shallow or rapid breathing
- Chest pain
- Muscle tension
- Blood pressure and circulatory changes
- Cold hands or feet
- Blushing, redness
- Trouble sleeping, nightmares
- Trembling, shaking
- Crying
- Loss of motivation
- Pelvic pain, vulvodynia

But despite how normal it is for our emotions to express themselves physically, "somatic" has somehow become a bad word in medicine. Women and racial/ethnic minorities in particular are often told their pain is psychosomatic—a term that has become synonymous with "it's all in your head," or "you're crazy." But take heed: Emotions are somatic *by definition*. All of us somaticize in one way or another every time we feel something. We can't help it; it's simply how we're built. Every emotion we have triggers a biological cascade in the human body, resulting in neurochemical, hormonal, cardiovascular, respiratory, digestive, musculoskeletal, circulatory, immunological, and other physiological changes. This can cause all kinds of symptoms and sensations—including pain.

The unfortunate messaging that this is abnormal or shameful has led to

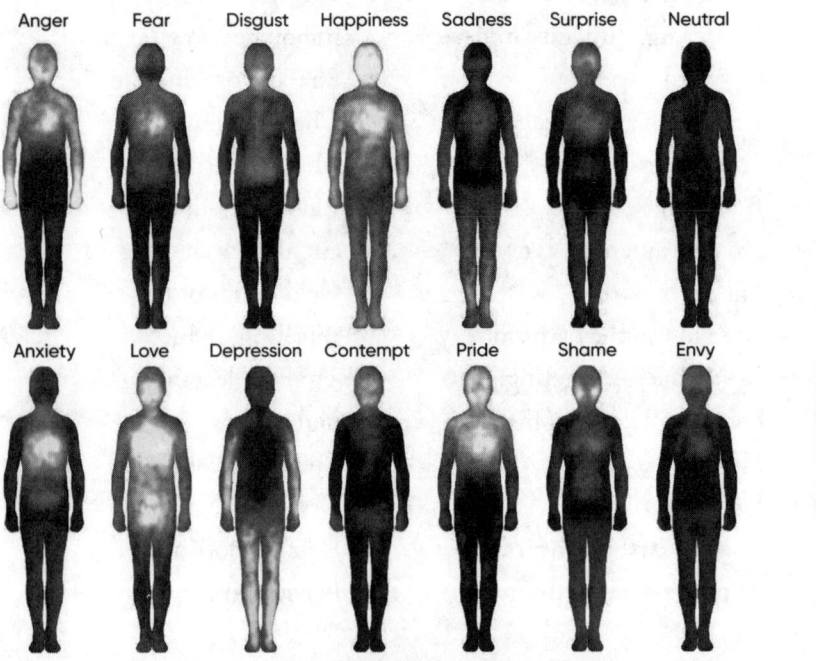

Topographical map indicating where different emotions are experienced in the human body.
Credit: Image from Nummenmaa L., Glerean E., Hari R. & Hietanen J. (2014), Bodily maps of emotions, Proceedings of the National Academy of Sciences, 111(2), 646–651.

massive misunderstanding, stigma, and the marginalization of millions. It has cast effective, nonpharmacological treatments that target emotional health into the shadowy pile of "alternative" or "complementary" medicine. So once and for all, let's get clear: Emotional health cannot, not ever, be separated from physical health. Because emotions are somatic, they're baked into every pain recipe—whether we like it or not. Targeting emotional health as part of chronic pain treatment is therefore neither alternative nor complementary: It is simply *medicine*.

Suppressed and Repressed Emotions: The Science of Stuffing It Down

That emotions are physical has serious repercussions for our health. Because negative emotions live in our bodies, they need a place to go. Consider this: If a teakettle didn't have an opening for the steam to come out, it would explode from the pressure. Humans are much the same. This is why we feel so much better after venting to friends, having a "good cry," or talking things out. But many of us deliberately suppress or unconsciously repress our emotions. While common, this can be detrimental to our well-being. Because if emotions aren't given a healthy release valve, they'll find ways to come out anyway.

Take Scott, for example. A broad-shouldered former athlete in his late twenties, Scott had missed most of college due to cluster headaches that felt like "ice picks through the head." No medication had helped. Deemed a medical mystery, he was referred to me. As he recounted his long struggle, tears quietly rolled down his cheeks. When I commented that pain seemed to have taken a lot from him, he nodded. But despite his tears, Scott always described his mood as "fine." The next few sessions were much the same: Scott would enter my office, sit on the couch, and immediately, and silently, start to cry. When I finally asked about his sadness, his blue eyes widened with surprise. "You mean this?" he said, swiping at his tears with disbelief and annoyance. "No, no," Scott said reassuringly, his face cracking into a wide smile as the tears continued to fall. "I'm not crying. I'm not even feeling sad. This is just something my face is doing."

We don't always suppress emotions on purpose; as with Scott, it can be inadvertent—an unconscious by-product of our culture, comfort levels, or upbringing. In many cultures, men are shamed into hiding their emotions and told from a young age that "boys don't cry." Our childhoods shape emotion expression, too. If there was never any room for us to express our feelings growing up, either because our culture didn't permit it or other family members took up all the emotional space, we might learn to shove them down. If it wasn't safe for us to express ourselves at home, it may have been adaptive to hide our emotions, or keep them to ourselves.

Regardless of the reason, emotion suppression is extraordinarily common. We've all seen people "explode" with unexpressed emotions: blowing up with rage, breaking objects in anger, hurting people, even becoming sick with unexpressed grief. These suppressed negative emotions come out in all sorts of other ways, too—muscle tension, vomiting, illness…and pain.

We now know that suppressed and repressed emotions can trigger physical symptoms, exacerbate illness, and amplify pain.[1,2] Emotion suppression may even increase our risk of early death. Identifying emotions and expressing them in healthy ways is therefore a critical part of the healing process.

Shared Neural Real Estate

So, you say, emotions are physical. But does that prove that pain is emotional…?

There's a good reason why we describe pain as "miserable" and "agonizing." Evolutionarily, the more unpleasant pain was, the more motivated our ancestors were to avoid danger, seek safety, and get help—and, thus, the more likely they were to survive. That's still true today: Pain's unpleasantness continues to be the very thing that saves our lives. The emotional component of pain is so fundamental to the pain experience that it even has its own name: In neuroscience, it is called the affective, or emotional, dimension of pain.

A look inside the brain offers a concrete explanation for why emotional pain and physical pain are so closely interconnected—why J. M. Barrie's health was so severely impacted by stress and grief, and why heartbreak can actually break our hearts. First, as we've now seen, is the fact that emotions express themselves physically. The second critical data point is that the parts of the brain that process emotion *also process pain*.[3,4,5,6]

Peeking inside the human skull with functional magnetic resonance imaging (fMRI), neuroscientists have been able to determine that pain and emotions share real estate in the brain, and are processed via overlapping brain networks and structures. There is such significant overlap, in fact, that many scientists consider the line between "physical pain" and "emotional pain" to be an extremely blurry one.[i]

These shared regions of brain overlap include the very structures and circuits we saw in Chapter 2—for example, the amygdala, an almond-shaped brain structure famous for its role in processing emotions like threat, fear, and anxiety. If input is potentially dangerous and threatening—a snake bite, for example—the amygdala cranks up pain volume. If input is unimportant and nonthreatening, however, like a harmless scratch, the amygdala turns pain volume down. A less-familiar site of overlap is the anterior cingulate cortex (ACC), which helps process pain's unpleasantness, or how much suffering and distress it causes. And then there's the insula, Latin for island, involved in emotional control of pain, empathic pain (feeling other people's pain), and interoceptive awareness—that is, our ability to know how our body is doing and feeling in any given moment.

The implication of this science is revolutionary. It means that every single sensory message from the body, whether it originates in the toe or the tailbone, must filter through the brain's emotion machinery *before* it can become the thing we call pain. In fact, pain couldn't be pain—at least, not the way we experience it—without the influence of emotions.

Sarah showed me this firsthand.

---

i The idea that pain consists of two completely discrete dimensions—sensory and affective—that can somehow be completely untangled is controversial. We experience pain as a single, unified entity, and it is only when these ingredients *come together* that they create pain. Despite controversies, experts agree that physical and emotional pain are intimately and fundamentally intertwined.

Stomachaches, Serotonin, and Stigma

Sarah, who wore thick glasses and a pixie haircut, looked younger than her forty-nine years. She was an established wildlife biologist at a local university and a single mother of two young girls. She had a deep, gravelly voice and always had a new word to teach me. ("Lagomorph" was my favorite: plant-eating mammals with two pairs of incisor teeth, like rabbits and hares). But that wasn't why she came to see me.

Sarah had suffered from debilitating stomachaches her entire life. They came with all manner of embarrassing gastrointestinal symptoms: diarrhea, nausea, gas, bloating, and occasional vomiting. She wasn't comfortable unless she was near a bathroom, which made traveling on public transit impossible and long car rides prohibitive. Over the years, her illness prevented her from going to work, attending social events, and sometimes leaving the house. Healthcare providers had blamed gluten, lactose intolerance, her gut biome, abdominal migraines, even a "fat allergy." Finally, she was diagnosed with irritable bowel syndrome, a chronic condition affecting the gastrointestinal (GI) tract characterized by disrupted gut-brain interactions. But despite dozens of doctor's appointments, elimination diets, and treatments, nothing helped. Sarah felt broken, untreatable, and ashamed. With all her training in biology, why wasn't she able to understand her own body? After exhausting all other options, she found her way to me.

Sarah disclosed that, as a child, she'd hidden behind her mother's skirts during social interactions, missed school due to stomachaches, got nauseous before tests, and had once fainted in Spanish class while giving a presentation. As she got older, she tended to have pain flares before parties, dates, and Sunday nights before work. More recently, public speaking, crowds, and public transit triggered symptoms. There was a pattern to her pain. But the medical exams had never assessed emotional health, and none of her providers had ever asked. She'd never been to therapy, and neither had her parents.

When I asked Sarah if she experienced any social anxiety, she requested a list of symptoms, considering each thoughtfully as I read them aloud. Persistent, intense fear of social situations that disrupts functioning? Check.

Heart palpitations and excessive sweating? "Hyperhidrosis," she said, grimacing and nodding in agreement. Urge to escape or avoid, check check. Among other things, Sarah doggedly avoided back-to-school nights, her daughters' plays, and piano recitals because of the anxiety triggered by meeting new people and being trapped in a room. She'd also developed a phobia of traveling on public transit as a result of her GI symptoms. Sarah met diagnostic criteria. But what, Sarah wondered aloud, did her mental health have to do with her abdominal pain? Her expression looked pained. Did this mean her pain was just psychological?

Decidedly not. Stress and anxiety affect the functioning of our GI tract, I explained, triggering symptoms like stomachaches and nausea. Why? Well, many of us have heard of serotonin: a brain chemical implicated in mood, and the popular target of many psychiatric drugs for anxiety and depression. But serotonin doesn't just live in our brain. *Ninety percent* of the body's serotonin lives in our gut. Specifically, it can be found in our enteric nervous system (ENS), a bundle of nerves connecting brain and gut. This ENS, also known as our second brain, helps explain why we have a "gut instinct," why nervousness causes "butterflies" in our stomach, why unkind words feel like a "punch in the gut," and why watching the news can make us "sick to our stomach."

Despite this, healthcare providers and patients alike are rarely told that the gut is a major part of the body's emotion system. It is easily dysregulated in times of stress—say, during an argument, after getting upsetting news, or in the face of major deadlines. It's no wonder one of the most common signs of stress and anxiety, in children and adults alike, is a stomachache. But until Sarah arrived at my office, she didn't know that her symptoms were linked to stress and anxiety, nor that emotions could trigger and exacerbate pain flares.

To reduce her stress, anxiety, and pain—all of them, together—we began a multipronged pain protocol. It included graded exposure, which entailed gradually exposing Sarah to tiny bits of feared stimuli. For example, the mere thought of riding the city bus gave her stomachaches and the sweats. Step 1 of our exposure plan was to walk to the bus stop, wait for the bus to arrive, and then leave. Step 2 was boarding the bus, riding it for just one stop, then getting off. After doing this for a few weeks, fear started abating—and Sarah's stomach mercifully stopped lurching.

In session, Sarah also learned strategies to soothe her nervous system and relax her gut, like diaphragmatic breathing, guided imagery, and mindfulness, and started using relaxation apps before bed. Practicing these helped rewire neural pathways that enabled her brain—and her gut—to access more relaxed states. As we tested out different strategies, Sarah discovered that touch and heat seemed to help, too; heating pads and hot water bottles reduced her abdominal pain and bloating, as did self-administered belly rubs.

We also made a list of life stressors that were contributing to her pain recipe, and started problem-solving. For example, because doomscrolling at night made sleep difficult, I recommended Sarah start a "social media diet" that involved temporarily deleting all social media apps from her phone. She limited news consumption to ten minutes in the morning and stopped watching it at night. She learned sleep hygiene strategies, including getting out of bed when sleepless instead of lying there becoming increasingly anxious about functioning the next day. She cut out dried fruits, carbonated beverages, and other foods that seemed to trigger flares. She began exercising at her local gym and lifting weights with a trainer. Every weekend, she drove to the woods for a long nature walk with her two girls.

Sarah also learned cognitive strategies to challenge anxious thoughts about flares, asking herself: "If the worst-case scenario happened, and I got sick in public, could I handle it?" She unexpectedly had the opportunity to test this when she vomited at a bus stop. Sarah was horrified and humiliated. But when the two kind souls who checked on her were utterly unfazed, the shame came and went. With it went her fear. The worst had happened, and she was okay. Once Sarah realized her predictions didn't match the facts, her fear of traveling on public transit diminished. She began carrying a plastic bag in her purse in case she got sick. Its presence was comforting: If the worst happened, at least now she was prepared.

Today, two years later, Sarah is still in touch. Many things are the same: She continues to teach me about wildlife biology. And to my delight, she also occasionally sends zoological words of the day. (Today's: "crepuscular," animals most active at dawn and dusk.) But some things are dramatically different. Sarah is sick less often, misses less work, and is in less pain. Her mood is better, and she is less anxious. She manages flares with confidence. Sarah has yet to miss a single recital, performance, or school event for her daughters. Some stressors, like long flights and presentations at work, still trigger flares. But learning to connect emotional health with physical health—recognizing that each is a fundamental part of the other—has given Sarah and her girls their beautiful life back.

Smart Phone, Smart Pain System

Remember that volume knob that lives in our central nervous system, constantly adjusting pain volume? This Pain Dial is where the relationship between pain and emotions takes center stage. Via a complex signaling system, negative emotions like stress and sadness magnify danger messages from the body, amplifying pain volume at the brain and spinal cord.[7,8,9] This results in a blaring, overprotective pain alarm. This is exactly what we saw with Sam in Chapter 1, whose body felt increasingly worse the more stressed, sad, and isolated he became. Similarly, Joyce's back pain intensified as she withdrew socially and dropped beloved hobbies.

There are benefits to having a "smart" pain system informed and adjusted by emotions. Negative emotions, like panic and fear, often arise when something dangerous or bad is happening. To ensure our body is prepared for this potential threat, emotions amplify pain to grab our attention and urge us to safety—whether that be by avoiding, guarding, resting, or seeking help.

However, *different* emotions affect the Pain Dial *differently*. Positive emotions like joy, gratitude, and relaxation cue our smart pain system to reduce pain volume. Why? Because these emotions typically occur when we're safe and secure: no need to waste precious resources on an alarm when no threat exists! In this way, positive emotions and mood states reduce—and can even inhibit—danger messages between body and brain, dialing down the pain alarm. This can mean less pain, and even a better ability to tolerate pain.[10,11,12] A team of Swiss researchers have shown that people are able to keep their hand in a bucket of freezing water longer if they're laughing at funny shows like *South Park* or *Friends*.[13] These same results have been found over and over again with different patients and age groups, from pediatric hospitals to nursing homes.[14,15] This phenomenon is known as positive-affect analgesia, which translates to "pain relief induced by positive emotions."

Positive emotions change our bodies in other useful ways, too. Joy, happiness, and gratitude increase brain levels of feel-good chemicals like serotonin, dopamine, and endorphins, our homemade painkillers. They reduce muscle tension, increase our energy, and even boost immune function. This is why we tend to feel better—both emotionally and physically—when engaged in pleasurable activities with friends, while relaxing on a beach vacation, or when we're simply in a good mood. When it comes to human health, it's actually true that laughter can be good medicine.[16,17,18]

This was true for Sam, whose brain started responding differently once his depression began lifting. This was true for Joyce, who started feeling better when she began cooking again, swimming, and experiencing more pleasure and joy. In fact, you'll witness this in action with every individual in this book—and have undoubtedly already experienced it yourself.

In this way, our amazing brains constantly adjust pain volume based on our emotional state: moderating and reducing it when we're safe, relaxed, and happy, and amplifying the alarm when we're fearful, distressed, angry, or sad. This doesn't mean that positive emotions "cure" pain (they don't), that happy people feel no pain (they definitely do), or that pain is solely the product of depression or stress (it isn't). We assuredly don't need to "just relax" or "be more positive." This type of dismissive, toxic positivity isn't true and doesn't help. What does help facilitate pain relief is recognizing that pain is always physical *and* emotional—never just one or the other, but always both. In fact, interventions and treatments that change our emotions can also change the pain we feel.

To see how this works, we need only pick an emotion and examine how it affects our pain system. And there is perhaps no better example of how an emotion adjusts the Pain Dial than stress.

Stress and the Superhero

A constant, dull hum in the background, our stressors come in seemingly countless forms: work, finances, family conflict, aging parents, an illness. Because stress is additive, these can pile in a mountainous mess and, without intervention, wreak havoc on the human body.

Our stress system is designed to respond to immediate, short-term threats, like a lion hunting us on the plains. But modern society has outstripped evolution, bringing with it ongoing stressors—screen addiction, mortgage payments, divisive politics, overheating oceans—that hijack our nervous system and leave us in a constant state of hyperarousal. Add chronic pain to this mix, and the combination is combustive. Because just as moving cross-country is a stressor, divorce is a stressor, and death of a loved one is a stressor, chronic pain is also a stressor—one of the most significant a human can endure.[19]

As you might suspect, chronic stress has consequences for both our emotional health and our physical health: It is a notorious pain trigger and amplifier.[20,21,22,23] This is because chronic stress affects all systems of the human body—triggering inflammation, tensing our muscles, suppressing our immune system, and sensitizing and amplifying the pain system. (It's no coincidence that muscle relaxants, massages, and stress-reduction techniques like mindfulness are all effective pain reduction strategies!)

However, in certain life-threatening situations, stress can actually do the opposite and *mute* pain. It can even give us superpowers. This is because of the way environment and social context constantly interact with emotions and biology—yet another example of the interconnectedness of the three domains of pain. This adaptation can increase our chances our survival. In fact, Mario's very life depended upon it.

Mario, age three, left his house and wandered onto the street, unseen by his parents.[24] The driver of the old Honda idling at the curb didn't notice him, either. The car suddenly stalled and rolled backward, pinning

Mario beneath. The child's neighbor, Michael, saw the commotion from his yard and realized Mario was trapped. He leaped up and ran toward the stalled vehicle. Using his bare hands, Michael lifted the three-thousand-pound car ten inches into the air—high enough for the child to be pulled to safety. Though he wore no gloves and had no help, Michael felt no pain. He later described the shocking surge of strength he'd felt to news cameras, which accompanied him back to the car for a repeat performance. But when Michael tried to lift the car a second time, he found the feat physically impossible: He couldn't even get the tires off the ground.

In the face of an urgent, acute stressor, our body temporarily suppresses the pain alarm, allowing us to fight, flee—and survive. This state of fight-or-flight triggered by stress prepares us to run for our lives, fight the threat, or freeze and play dead. As a result, the brain releases neurochemicals that temporarily lower pain volume, like homemade opioids, while hormones like adrenaline and noradrenaline liberate energy and power to give us superhuman strength.[25] This is adaptive, because if we're injured during that emergency—by the lion trying to eat us, the stranger attacking us, the car pinning our child—pain won't stop us from running, lifting, or fighting to stay alive. This phenomenon is known as stress-induced analgesia, and, thanks to the way emotions change pain, *all* of us possess this superpower.

Mental Health IS Physical Health: Anxiety, Depression, and Pain

Given the powerful ways stress interacts with pain, it shouldn't surprise us that anxiety and depression do, too. Anxiety and depression symptoms are five times more common in people who have chronic pain than in those who don't.[26] Anxiety, which triggers an intensified, chronic fight-or-flight response, comes with its own host of physical symptoms, some of which you've likely experienced: feeling restless and on edge, racing heart, muscle tension, headaches, stomachaches, nausea, shallow or rapid breathing, an urge to flee or escape, and insomnia. In fact, among the most common signs of anxiety in kids are headaches and stomachaches. Like stress, anxiety also puts our brain in threat mode, which cranks up the Pain

Dial.[27,28,29] And the more anxious we get, the more our muscles tense, the more reactive our pain system becomes, and the more we hurt.

Should we need any further evidence that mental health is connected to physical health, a relationship has been found between depression and pain, too. Anyone living with depression will tell you that depression actually hurts. That's because depression also shares neural networks with pain, increasing pain sensitivity while reducing pain tolerance.[30] Depression may be classified as a "mental health" issue, but it's also physical—commonly accompanied by a host of bodily changes. Those living with it may experience crushing fatigue, disrupted sleep and appetite, loss of libido, joint and body pain, and a heaviness that can make something as simple as getting off the couch feel impossible.

This creates a well-known cycle, one we've seen before: As emotional suffering increases, so too does pain. And the worse our pain, the more depressed, stressed, and anxious we're likely to become. Around and around we go, physical and emotional pain antagonizing each other in a brutal, vicious cycle.

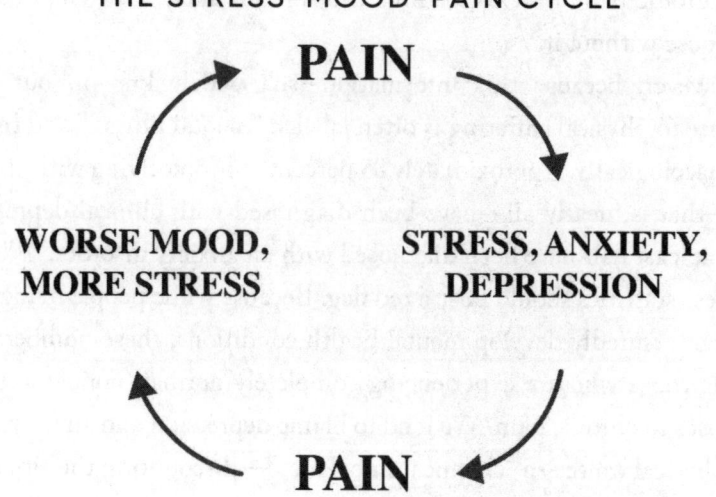

Credit: Illustration by Eva Huzella

Unsurprisingly, the opposite is true, too: When we feel better emotionally, we tend to feel better physically. Ultimately, whether the pain triggered the depression or the depression triggered the pain hardly matters—what matters is that our emotions have the power to trigger, amplify, *and also reduce pain* at every level of the pain experience.

A Normal Response to an Abnormal Situation

It's important to note that experiencing some measure of anxiety and depression when we have chronic pain is completely normal—even expected. Why...? Because, as Sarah, Sam, Fallon, and Joyce all experienced, chronic pain is a thief, stealing our ability to work, play, and engage in beloved hobbies. It can reduce our life to a shadow of what it was before. Over time, this naturally triggers grief, depression, and fear—immense sadness about the present, and immense anxiety about the future. *This is a normal response to an abnormal situation.* The human body wasn't built to be in pain day in and day out for years on end. Research bears this out: The more pain we have, the more anxious, depressed, and hopeless we're likely to feel. In some cases, this hopelessness can become lethal. People living with chronic pain are two to three times more likely to attempt suicide than those without it.[31]

However, because this information isn't widely known, our basic response to physical suffering is often labeled "mental illness," and treated pharmacologically. Approximately 85 percent of people living with chronic pain—that is, nearly all—have been diagnosed with clinical depression, while at least half have been diagnosed with an anxiety disorder.[32,33]

These statistics should raise a red flag. Because while people living with pain can assuredly develop mental health conditions, these numbers also include those who are experiencing completely normal, nonpathological responses to chronic pain. We tend to blame depression and anxiety solely on a physical cause—a "chemical imbalance"—discounting the situations and circumstances surrounding them. But environment plays a massive role, too. Despite what Pharma marketing campaigns proclaim, anxiety and depression are *never* purely chemical: As with pain, mental health is biopsychosocial, too. When pain is the primary driver, these mood states are considered situational: situational anxiety, and situational depression. Indeed, for the vast majority of my patients, when the pain resolves, the anxiety and depression do, too.

This is extremely hopeful news—because anxiety and depression are treatable, and stress is manageable. There are a million ways to cultivate positive emotions and decrease negative ones—methods that change both our psychology *and* our biology. You will learn how to access this Healing Cycle of positive emotions and pain reduction in Part III.

If stress can slow our growth and anxiety can sicken our stomachs—if emotions can adjust our Pain Dial, acting as amplifiers and attenuators—the story of pain *must* be deeper than the one we've been sold. Pain, it turns out, is an iceberg: physical symptoms above the water, visible for all to see; and countless other factors, like emotions, hidden below. And if emotions are inseparable from how pain is made, this makes them inseparable from how it's healed.

The next pillar of pain is even more invisible. Science shows that pain is also *cognitive*, shaped and influenced by thoughts, predictions, beliefs, and where we place our attention. But because these events occur inside the confines of our skull, it's typically impossible to observe what they do to our bodies.

Until now.

6

Pain Is Cognitive

THE MIND, THE BRAIN, AND PAIN

Thoughts Change Pain

In a tiny Himalayan village atop a snowy mountain, Tibetan Buddhist monks sit quietly in a cold room. It's forty degrees Fahrenheit, hovering close to freezing. Cold, wet sheets are draped across their bare shoulders—conditions that would typically risk causing hypothermia. The monks close their eyes and begin the sacred practice of Tummo meditation, otherwise known as the "yoga of inner heat." As they meditate, they visualize air entering their lungs and transforming into energy and warmth. They focus on images of fire in their chest, radiating throughout their bodies, warming their core and limbs. The monks do not shiver. Instead, steam begins to rise from the sheets. Sensors affixed to their skin confirm that they are generating enough heat not only to warm themselves—but also to dry the cold, wet sheets on their backs in under an hour. Tummo is practiced in even harsher environments during training periods: outdoors, in the snow, in temperatures as low as thirty degrees below zero. The monks wear only thin ritual garb, meditating throughout the frigid night until the sun rises, while the ice around them melts.

Is this some sort of ancient power?

Perhaps, in part. But it is also readily explained by science.

The power of thoughts and imagery to change our bodies has been known for centuries, evidenced across religions and cultures. Astronauts use it to prepare for space flight. Olympians use it to visualize and achieve perfect form during high-stakes competitions. And Alex Honnold, record-breaking climber, used visualization as a tool to help him scamper up El Capitan. But *how* do our thoughts exert these effects on our bodies?

We've been trained to believe that thoughts occur exclusively in the space between our ears. But thoughts aren't just bubbles that appear above our heads like they do in cartoons. Rather, thoughts are physical: electrical, neurochemical events that trigger a cascade of biological events throughout the body. If you've ever awoken sweaty from a nightmare with your heart racing, had tears spring to your eyes at the memory of a deceased loved one, salivated while imagining your favorite meal, or had an erotic fantasy and become physically aroused, you've experienced firsthand the incredible power of cognitions over your body.[i]

Over the past few decades, we've seen an explosion of knowledge about how the mind and brain interact with the body. There are now entire branches of science dedicated to these studies, including psychoneuroimmunology: the study of how cognitions and emotions interact with our biological systems to impact health and immune functioning.[1] It shouldn't be a surprise, then, that thoughts can do things like adjust our body temperature. But it certainly was to me.

Many years ago, when I first opened my private practice for people living with pain, I became intimately familiar with this connection between thoughts and the body. "Do you use biofeedback?" asked a colleague. I'd heard of it, but had never studied it, so immediately sought out a biofeedback specialist for training. I found Dr. Erik Peper, an acclaimed biofeedback expert and professor at San Francisco State University. Within minutes of meeting, he asked if I wanted to learn how to warm my hands to ninety degrees.

---

[i] Don't buy the BS story that impotence, pelvic pain, vulvodynia, and other forms of "sexual dysfunction" are entirely biological. *You aren't dysfunctional.* Like everything to do with human health, sexual issues are biopsychosocial—influenced as much by our thoughts, emotions, and environment as by hormones, blood flow, and biochemistry.

I stared at him in disbelief, the neuroscientist in me wondering if this was a sham. "With all due respect, Doctor," I said skeptically, "I don't believe in magic." A chronically cold-handed person, I didn't know until that moment that my cold extremities were a sign of stress. (It had never occurred to me that "getting cold feet" before a wedding might be literal, let alone rooted in neuroscience.) Dr. Peper—an undeniably great name for a doctor—explained the science while hooking me up to his intricate-looking biofeedback machine. Negative thoughts and images, he said, activate the body's sympathetic stress system, resulting in muscle tension, reduced blood flow to extremities, and, ultimately, cold hands and feet. Positive, calming thoughts and images, he said, would do the opposite: reduce muscle tension, increase circulation to my extremities, and increase skin temperature.

In my thirty-plus years of research and training, I'd never heard of such a thing. So this remained to be seen.

Dr. Peper attached sensors to my fingertips, abdomen, wrists, and rib cage, explaining that the sensors would provide instant feedback about various biological processes: heart rate, galvanic skin response, respiration, and skin temperature. ("Lie detector" devices are in fact biofeedback machines that detect the physiological changes that accompany the stress of lying.) He then instructed me to think stressful thoughts. I needed no prompting: Images of endless to-do lists, unpaid bills, familial obligations, and the faces of struggling patients appeared before me. My heart rate instantly escalated, my muscle tension spiked, and my skin temperature plummeted. The numbers were plainly visible on the screen before me. The more stressed I got, the colder my hands became. While I couldn't detect these small changes, the machine could. Cognitive processes, Dr. Peper explained, always, and powerfully, change the body—even if we don't notice it happening.

He then guided me through some practices that will soon be familiar to you: diaphragmatic breathing, relaxation, and guided imagery. To my surprise, within minutes, my hands began to warm. At first, I thought I must be imagining it. But when I opened my eyes to peek at the screen, the numbers were steadily increasing. And when I looked at my hands, my palms were turning red. Within a few sessions, I was able to warm my hands to ninety degrees. It's gotten easier, and now happens faster, with practice and time. Recently, on a hike in the winter snow, I practiced my hand-warming skills to see if I could do it under pressure. After a few minutes of intense concentration, I was the only person on the hike not wearing gloves.

My hands were warm.

The Pain Cycle

Given the evidence we've seen so far, it shouldn't surprise you to learn that thoughts also have a profound impact on pain.[2,3,4] In fact, it is our very thoughts that create and perpetuate something that I call the Pain Cycle.

This cycle is perhaps the most fundamentally important one you'll learn about in this book—because it plays a key role in how pain is made, and also how it heals. The Pain Cycle consists of four main elements: thoughts, emotions, physical sensations, and behaviors. These inextricably linked components always affect one other: The things we think impact how we feel physically and emotionally, as well as the behaviors we ultimately choose. And this entire cycle also works in reverse.

The Pain Cycle begins with a simple thought, like this familiar one: *I'm broken, I'll never get better.* This belief of brokenness triggers feelings of sadness, fear, and hopelessness. These negative emotions, in turn, trigger a biological cascade of changes across bodily systems: neurochemical, musculoskeletal, immunological, digestive, cardiac, circulatory, respiratory, and endocrine. *I'm broken, I'll never get better*: Stress hormones spike, immune function is suppressed. The brain produces fewer feel-good neurochemicals and natural painkillers. Muscle tension increases, and sleep is disrupted. *I'm broken, I'll never get better*: Our central nervous system sensitizes and the brain's pain alarm amplifies, making our body feel even worse than it did before.

This combination of thoughts, emotions, and physical sensations then inspires us to act, or behave, in certain ways. These behaviors are our attempts to cope with pain and illness. A coping behavior can be something we do—like going to the doctor—and it can also be something we stop doing. When it comes to pain, this often includes stopping activities, withdrawing from work and play, and avoiding movement. But this avoidance, paradoxically, makes us worse. It results in muscle atrophy, stiffer joints and muscles, and reduced strength and mobility. Ultimately, these coping behaviors trigger increased disability, a more sensitive pain system, and more pain.[5]

It doesn't end there. Our behaviors and daily choices then circle back around to affect our mood and our mindset. The more we withdraw, avoid, and immobilize, the darker our thoughts become, the more negative our emotions, and the worse we ultimately feel. As we fall behind socially, lag professionally, and become more isolated, stress and anxiety spike and mood crashes. As a result, the Pain Dial turns up, *making pain even worse*. Thus, the Pain Cycle perpetuates with little to stop it.

We saw this with Fallon, the bed-bound painter with complex regional pain syndrome (CRPS), also known as the "suicide disease." After countless failed treatments, her belief that she was beyond help triggered fear and hopelessness. These negative emotions resulted in neurochemical and physical changes that amplified pain and impaired functioning. As a result, Fallon coped by staying home and minimizing movement. She stopped going to work, stopped seeing friends, and quit archery. Stripped of social support, joyful activities, and a normal life, Fallon's pain worsened—and her thoughts and emotions got even darker.

This cycle kept Fallon in bed for three long years, until the day she broke it, and finally began to heal. Increasing agency and hope, getting outside and moving, interrupting self-defeating thoughts, spending time with friends, improving emotional health, reducing stressors, improving sleep, and other interventions changed Fallon's body, her brain—and her pain.

This isn't to say that we can "think our way out of pain"—it isn't that simple. Nor is this a suggestion to just "think happy thoughts." That isn't

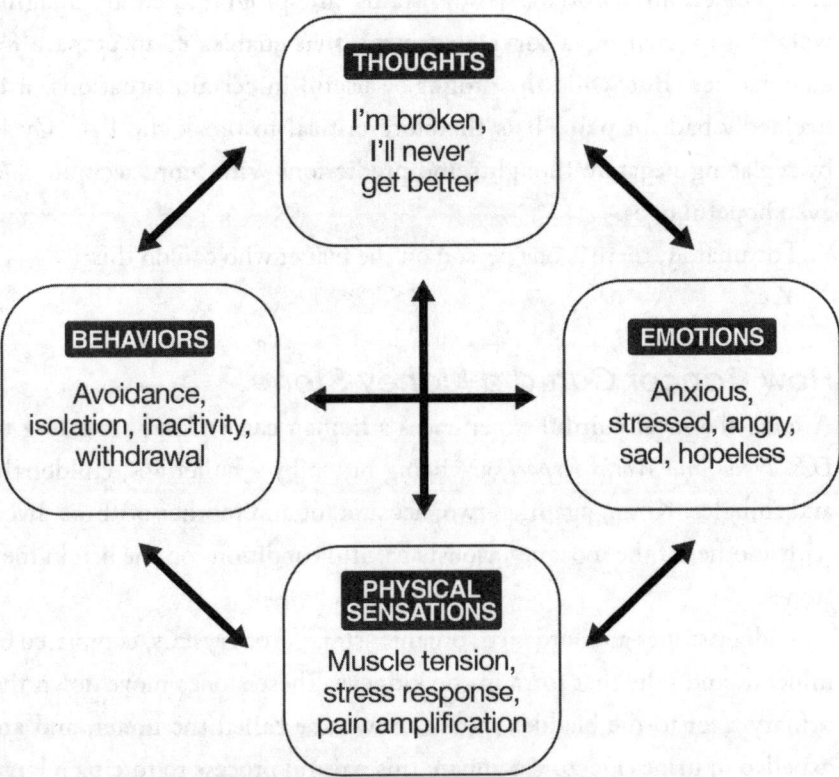

Thoughts, emotions, physical sensations, and behaviors constantly interact to change the pain we feel.

Credit: Illustration by Eva Huzella

realistic, and toxic positivity has no place in medicine. However, pain is aversive and scary by definition, inspiring fearful, untrue thoughts and predictions. Moreover, our brains are programmed to imagine worst-case scenarios, a survival strategy that enables us to prepare for emergencies. But while this might be useful in certain situations, it is decidedly bad for pain. It is therefore critical to break the Pain Cycle by replacing negative thoughts and predictions with more accurate and even hopeful ones.

Fortunately, there is one person on the planet who can do this:
You.

How Cancer Cured a Kidney Stone

Among the most painful experiences a human can endure, according to *U.S. News and World Report*, are being bitten by a bullet ant, childbirth, and shingles. Kiran, age fifty-two, accountant and mother of three, lived with another of the most notoriously painful conditions on the list: kidney stones.

Kidney stones are hardened, organic "stones," or crystals, comprised of minerals and salts that form in the kidneys. These stones move down the urinary tract to the bladder via a narrow tube called the ureter, and are expelled in urine. Doctors compare this painful process to forcing a large pebble through a teeny, tiny straw.

Kiran was generally healthy, she told me when I first met her, but her family had been plagued by health issues in recent years. Earlier that year, her still-young, beloved father had been diagnosed with stomach cancer. He'd fought a good fight, but in the end, the cancer won. They'd lost him just a few months before. It had been a shock to the whole family—but it rocked Kiran, his closest confidant, most of all.

Kiran's kidney stones seemed benign in comparison; she had a high pain tolerance, she said, and rarely complained. She always knew when she had developed one from the telltale symptoms she felt a few times each year: low back pain, abdominal pain, nausea, and frequent urge to urinate. The stones always took about two weeks to pass from symptom onset—sometimes less, but never more. The pain was generally the same, too: approximately 6 out of 10 on the pain scale. Her family doctor recommended that Kiran manage the pain by taking her prescribed medications, drinking copious amounts of water, and distracting her brain as much as possible. After doing some reading, Kiran also changed her diet, ate less salt, eschewed calcium supplements, and even became a vegetarian. Her efforts definitely helped. But still, the stones appeared.

One spring morning, Kiran was experiencing the usual abdominal pain that typically accompanied her kidney stones, a 6 out of 10 on the pain scale. She'd taken her medication and used her other strategies, but the pain remained the same. When she winced, her husband looked up from his desk, alarmed. "More pain?" he asked, his eyebrows raised in concern. "It's been an entire month since your symptoms started. Are you sure this is a stone, and not...something else?"

Kiran froze. A month? That wasn't normal. Panic rose in her chest. Her pulse quickened. She thought, *This isn't a kidney stone. This is cancer, the same stomach cancer that killed my dad.* Vivid images flashed before her: the ominous, shadowy outline of the growing mass visible on her father's abdominal scans. His pale, drawn face and hollow cheeks. His sunken eyes, tight with pain. Her pain shot to an 11 out of 10 on the pain scale. She doubled over and fell to the floor. When she cried out, her husband called 911, and Kiran was rushed to the emergency room.

When she arrived at the hospital, the doctor performed an abdominal scan. Kiran impatiently watched the monitor as they searched for the culprit. She could see her stomach, intestines, bladder, kidneys...and suddenly, there it was, plain as day, illuminated on the screen: a kidney stone. Her doctor confirmed that the stone was indeed the likely source of her pain, and that, given its location and the lack of other concerning symptoms, it would likely pass in a matter of days. The doctor also said that she could find zero evidence of cancer, and that Kiran was safe to go home. As soon as she heard this, Kiran's pain instantaneously and inexplicably dropped to a 3 out of 10. She and her husband left the hospital, got in their car, and drove home. As predicted, the stone passed in a matter of days; during that time, Kiran's pain never went above the usual 6 out of 10.

If Kiran hadn't experienced it herself, she later told me, she never would've believed it was possible. It was her *thoughts*—electrical, chemical messages—that had turned up her Pain Dial. Not a shifting stone, not a sudden tumor. Delivering new imagery and messages of safety to her brain was, in Kiran's case, the missing medicine.

Meet Pain Voice

Now is the perfect time to introduce you to an old friend: your inner dialogue, the voice of worries, fears, negative self-talk, what-ifs, and worst-case scenarios. Because of the impact pain can have on our self-talk, I've named this voice Pain Voice. Pain Voice feeds us pessimistic, catastrophic, unhelpful thoughts that perpetuate the Pain Cycle and amplify pain. These are often future-oriented, and may sound something like this: *I'm broken, I'll never get better. Nothing has helped so far, so nothing will ever help. This pain is never going to end. Pain is going to ruin my life and relationships. My body hates me. I'm a burden to my friends and family.* Kiran's Pain Voice, for example, erroneously told her that her kidney stone was cancer, triggering panic and fear. Sam's Pain Voice told him he'd never play soccer again, let alone graduate from high school. Yours likely says different things to you.

These negative Pain Voice thoughts are commonly known as cognitive distortions—because while they might sound true, they're actually distorted, unhelpful, and false. Having these thoughts isn't unusual, especially when we have pain. In fact, it's expected. The brain is a prediction machine, and chronic pain changes those very predictions. When the brain gets stuck in danger mode, it's more likely to magnify and misinterpret stimuli, exaggerate the direness of symptoms, and predict terrible outcomes. A brain in pain thus has a tendency to make negative predictions about our bodies, our future, and our recovery—discounting and ignoring positive possibilities.[6,7]

WHAT A BRAIN IN PAIN PREDICTS

A brain in pain is in danger mode, making it prone to negative predictions.
Credit: Rachel Zoffness Instagram, @therealdoczoff

In fairness, to predict bad things and imagine worst-case scenarios is human—everyone does it. There's a reason for this: a negativity bias that privileges and prioritizes negative, dangerous input helped our ancestors survive. Pain is an aversive human experience, something we're built to fear and avoid. Having negative thoughts when we're in pain, therefore, is normal—not pathological—just like having negative emotions. It can often require Herculean effort to *not* catastrophize when we receive catastrophic news, like a scary diagnosis, or have frightening symptoms, like life-changing pain.

But because thoughts are physical and have biological consequences, Pain Voice can wreak havoc on our bodies. Negative predictions trigger the sympathetic stress response, crush hope and motivation, and increase pain and disability. This is exactly what we saw with Kiran, whose interlocked stress and pain systems launched into overdrive when Pain Voice proclaimed she had cancer. Pain Voice is even associated with changes in the very structure and function of our brain, activating and strengthening pain pathways.[8] Because of neuroplasticity, the more we repeat negative, pessimistic messages, the stronger the brain pathways dedicated to them become. All told, this negative self-talk can reduce our likelihood of recovery, crushing our mood and encouraging us to avoid the very activities we need in order to heal.

This doesn't mean that healing pain is merely "mind over matter," or that we can think our way out of pain. It just isn't that simple, and other ingredients matter, too. However, with awareness of Pain Voice comes great power. Because learning to recognize it—and challenge it—gives us yet another way to break the Pain Cycle.

Hope Floats

Lest we get discouraged, our bodies aren't affected just by negative thoughts. Positive thoughts change our bodies, too—only this time for the better.

When I was a little girl, my mother took me to see *Peter Pan* live on Broadway. The fairy tale, written by our grief-stricken friend J. M. Barrie, was captivating. Peter Pan, a little boy who can fly, has the ability to help other children fly, too. He has almost enough magic to lift them off the ground and guide them to Never Never Land, save for one final, missing ingredient: "Think lovely, wonderful thoughts," Peter instructs them, "and they will lift you up in the air." The eager children, grounded by gravity, generate powerfully joyous thoughts: candy, Christmas, sailing. The happier they become, the more buoyant they are—until, with a sprinkle of pixie dust, they are airborne.

This resonates with us because there's some truth to it. Hopeful, optimistic, joyful thoughts don't just lift our spirits and boost our mood; they also have an incredible impact on our biology. One study that compellingly demonstrated this was conducted at Johns Hopkins Medical School in the 1950s.

A scientist named Dr. Curt Richter placed wild rats in jars of water to examine their endurance. What factors, he wondered, determined the length of time they were willing—and able—to swim? He hypothesized that, in addition to things like physical strength, cognitive factors like hope and motivation played a key role. To test this, he restrained the rats, extinguishing all hope of escape. On average, the hopeless rats were able to swim for a maximum of fifteen minutes, after which they gave up and drowned.[ii] Some of them, imprisoned and terrified, lasted less than sixty seconds.

He tried this again with a new batch of rats. But on this next pass, just as the rats were becoming fatigued and starting to sink, he reached into the water, lifted them out, stroked them and fed them. Warmed and energized, the animals were then placed back into the water. He repeated this multiple times, teaching the rats that the situation wasn't actually hopeless—and that salvation was possible. When he put them back into the water for their final swim, the animals treaded water for up to *eighty-one hours*. Powered by the hope that a lifesaving hand would again appear, they swam for more than three full days, showing no signs of giving up. "After elimination of hopelessness," Dr. Richter wrote, "the rats do not die."[9] Hope, he concluded, and lack of it, so drastically changed the animals' physiology that it quite literally meant the difference between life and death.

As the saying goes: Hope floats.

Science suggests that humans are much the same. While "the power of positive thinking" might sound like pop psychology nonsense, research shows that, when it comes to our health, mindset really *does* matter. Hope, optimism, and a positive outlook can enhance immune functioning, reduce our risk of health conditions like cancer and cardiovascular disease, and increase our lifespan.[10] A positive mindset can boost feel-good neurotransmitters like serotonin and dopamine, and protect against

---

ii Dr. Richter's animal experiments were brutal. However, lest we judge him too harshly, this was standard protocol for scientific experiments at the time. Moreover, he'd been tasked with developing new ways to control Baltimore's overwhelming rat population. Source: Ramsden, E. (2011). Model organisms and model environments: A rodent laboratory in science, medicine, and society. *Medical History*, 55(3), 365–368.

mental health issues like depression and anxiety. Positive thoughts also have a substantial impact on pain.[11,12] In fact, 70 percent of studies on pain and optimism show a connection between hopeful thoughts and better health outcomes—including reductions in pain intensity, frequency, and disability.[13]

Changing our mindset doesn't just alter our neurochemistry and biology. It also changes our behavior. When we're hopeful about our recovery and optimistic that we can effect change, we're more likely to make choices that break the Pain Cycle—swapping unhealthy behaviors for healthier ones, moving our bodies, seeking social support, and eating better. Ultimately, our mindset helps determine how empowered we feel to find solutions, and how likely we are to implement them.

When I first met Mateo, for example, the child who'd lost his arm in an explosion, I knew that if I was going to have any chance of helping him, he first needed to *believe* that change was possible. So I injected him with a massive dose of hope, telling him that I could help him if he was willing to try my medicine—even though it wasn't in the shape of a pill. It was Mateo's belief in this plan, his willingness to have hope, that motivated him to try the treatments and strategies I offer you in this book. He tried physical and occupational therapy, met Emmett the Explorer, and used the strategies he learned in our sessions to improve his health and reduce his pain. Mateo's subsequent return to school and socializing—and his decision to shed the sweatshirt in which he'd been hiding—were powerful evidence that the choices he'd made had changed his life.

Clearly, it's not just rats that swim longer when given a dose of hope.

We do, too.

Pain Voice may be pervasive (and persuasive!), but our thoughts aren't fixed, dooming us to negativity for all of eternity. Rather, they are flexible and malleable. Indeed, we are *constantly* changing our own minds. If addressing this hasn't yet been part of your treatment protocol, we'll fix that soon enough. First, let's look at another important, oft-overlooked cognitive ingredient in our pain recipes: attention.

Attention Is a Magnifying Glass

If the thoughts that flow through our minds form a river, *attention*—another important cognitive ingredient in our pain recipe—is the current that directs them. The things we choose to focus on tend to dominate our cognitive real estate. Focus on a chess game, for example, and, for a brief time, the world will sharpen and shrink down to the pieces on that checkered board. Focus your attention on your hike, and your brain will fill with birdsong, fresh air, and blue sky. But it isn't just sights and scents that grow as we attend them. Sensations do, too.

In fact, attention—and its inverse, distraction—are the reasons why distracting a child from an injection makes it hurt less, why describing our back pain to our doctor makes it seem to hurt more, why we can focus on nothing else after burning our hand on the stove, and why we briefly "forget" our pain when absorbed in pleasurable activities. To illustrate how this works, let's enlist the help of a gorilla.

The Invisible Gorilla

In 1999, cognitive scientists at Harvard University conducted an experiment exploring how human perception is altered by attention and expectation.[14] Study participants watched a video of two teams passing basketballs back and forth. They were asked to selectively attend to only one team, ignoring the other, while counting how many times players on their chosen team passed the ball. At the end of the study, the observers answered a series of questions. One of these was: While you were counting, did you notice anything unusual?

Of the nearly two hundred observers, approximately half failed to notice the man in the hairy gorilla suit walking among the basketball players. Upon replaying the tape, the massive gorilla was so plainly obvious to the observers that they were gobsmacked to discover they'd missed it. In a revised version of the study using different observers, the researchers made the gorilla even more obvious. This time, it walked into the middle of the basketball game, turned to face the camera, thumped its chest, and then continued walking across the court. This event lasted a full nine seconds.

Again, half of the observers didn't even notice.

This quirk of human perception is called "inattentional blindness," and it reminds us that ultimately, it isn't the eyes that perceive: It's the *brain*. And what our brains perceive isn't solely dependent on the things our eyes see, the scents our noses smell, or the sensations our skin detects—it is also orchestrated by attention. Attention influences what incoming information is coded by the brain as noteworthy or important, or, conversely, completely ignored. If you watch the video from this experiment having read this book, it will be impossible *not* to attend to the gorilla. However, a naïve viewer asked to focus on ball bounces is likely to miss the animal interlude entirely.

Attention serves as the brain's magnifying lens, zooming in to amplify information for further examination, particularly if it's unfamiliar or potentially dangerous, and zooming out when input is safe, boring, or benign. Attention is an incredibly powerful tool for changing sensation, perception, and the way our bodies feel, making it *a critical part of every pain recipe*.[15,16] As with other cognitive factors, like the thoughts we think and the mindsets we adopt, attention is, at least to some degree, under our conscious control.

Kamal's Invisible Gorilla

Kamal, sixty-eight, a big-hearted retiree whose laugh filled rooms and reverberated off walls, was having a bad pain day. His rheumatoid arthritis, which he blamed on a lifetime of rough contact sports, had been acting up. Many things seemed to make his pain worse these days, including weather, poor sleep, and how much (or, more accurately, how little) he moved his body.

Over months of treatment, Kamal and I experimented with a variety of techniques to lower his pain volume. We identified several successful strategies, including heating pads and icy-hot creams, which soothed his joints and distracted his brain with new sensations; weekly massages; and guided meditations that led his mind to relaxing places in nature.

But the most powerful tool seemed to be distraction—which, he laughingly reported, came in a few reliable forms: *Sunday Night Football*, tinkering with projects in his garage, and playing with Bear, his giant brown mutt, who seemed to be as much teddy bear as he was dog. On a few memorable occasions, Kamal reported that he became so completely absorbed in his hobbies that he briefly, and entirely, forgot about his pain.

On this day, sitting in his living room, neither pain medications nor imagery exercises was working. So, Kamal moved to the next strategy on his list: distraction. He painfully limped into the yard with Bear at his heels, the dog's tail wagging joyfully. As Kamal began a game of fetch, enjoying the sensation of the warm sun on his skin, his pulse began to slow and his muscles relax. When Bear started tugging at the ball and playfully tackled him, Kamal began to laugh. Covered in slobber, laid out on the grass, he was suddenly flooded with memories of wrestling with the goofy Labrador retriever he'd so loved as a child. He left his body behind and traveled back in time. Lost in distractions, Kamal continued to play with Bear as the afternoon slid by. The shadows grew long. He barely noticed.

Hours later, Kamal's son sauntered outside. "Hey, Dad!" he shouted good-naturedly. "How's your pain today?"

Suddenly, it was as if Kamal was snatched from the clouds and slammed painfully back to earth, trapped again in his aging body. All of his attention went rushing back to his joints and the places that hurt. He looked around, suddenly aware of the late hour, his rumbling stomach, and his aching body.

Defeated, Kamal answered, "I was at a three, but now I'm at a nine."

The Neuroscience of Attention

In our sensory-rich environments, our brains are constantly bombarded by competing inputs, all clamoring for our attention—hungry kids, pinging texts, incoming sensory messages from our body. If our brains didn't have a way to filter out unimportant data, to separate meaning from noise, we'd constantly waste precious energy on useless information, like the sensation of clothes on our skin. Attending equally to every sensation would create a chaotic cacophony, overwhelming our brains and impairing functioning. To conserve valuable resources, therefore, our bodies developed mechanisms to automatically—and deliberately—filter out the seemingly useless, and dial in to the potentially important.

How the brain automatically zooms in and out without our executive say-so depends upon many factors, including the type of stimulus vying for our attention. For example, a stimulus that is "deviant"—that is, novel, unexpected, or unfamiliar—captures our attention because it is potentially noteworthy, such as an alarm that pierces a silent night, or a sharp pain where before there was none. These are examples of the involuntary capture of attention, or bottom-up attentional processing, wherein the brain automatically focuses on stimuli without our consciously asking it to.[17] The brain naturally attends to differentness, because it may convey important information about our safety and well-being.

Our brains also automatically attend to "salient" stimuli—that is, relevant and important—depending upon how critical or dangerous our brains perceive it to be. Failure to attend to sudden pain could mean death if it's due to a poisonous spider bite, for example. These two qualities, deviant and salient, help explain *why* the brain zooms in when it does. In fact, this is exactly why pain is so difficult to ignore: Pain is both deviant, differing from the norm, and salient, because it can portend potential danger.

These same rules apply to stimuli that activate any of our five senses, including touch. We've all seen how the brain dials in and out to magnify or minimize sensations: the way a new piece of jewelry, like a wedding band, feels foreign and strange on our hand—until, one day, it simply becomes part of the body's landscape, the brain ignoring it so completely that we forget it's even there. Take it off again, and your finger will feel oddly naked. If you've ever searched for your lost sunglasses, comically finding them still on your face; fallen asleep with contacts in, completely forgetting they were there; or stopped noticing the deafening roar of an airplane engine during a flight, attention science explains why. It's also why, at night, when we close our eyes and no longer have a million inputs competing for our attention, our pain suddenly becomes deafening.

Indeed, this is the great irony: Despite the fact that pain demands our attention, an ongoing, telescopic focus on the painful body part doesn't help—it actually hurts. In fact, the more we attend to pain, the higher pain volume climbs.[18]

Taking Control: The Power of Distraction

Luckily, conscious control—that is, where we *want* to place our attention—matters, too. If we ask our brain to focus on something, even if it's not particularly dangerous or important, it will. This is called "selective attention," or "top-down" attentional processing, meaning we get to select the objects of our attention according to our goals and motivations.

As evidence, try this: Wherever you are, stop for a moment. Close your eyes. Tune in to everything you can hear: the cars on the street, a barking dog, the wind in the trees. Really focus on those sounds until you've gathered all the details. Did you perceive any sounds you hadn't noticed before? Now, take all of that attention and drive it inward, into your body. Notice if your eyes are strained or soft, your shoulders hunched, your back slouched. As you focus your attention inward, are you more aware of sensations? Do any seem more exaggerated than before? When I did this exercise, I discovered that my jaw muscles were so tight that I had to consciously unclench them. Hey presto, *my* invisible gorilla: important sensory information, drowned out by other inputs, suddenly obvious when I attended to it. In this way, painful stimuli can disappear, reappear, and disappear again depending upon our focus.

We can also actively, deliberately zoom out: Anytime you've decided to ignore your phone and focus on your kids instead, or concentrated so completely on a hobby that you briefly "forgot" about your pain, you've relied upon this mechanism to do it. This proves to be an important tool we can quickly use to lower pain volume.[19] Virtual reality, thermal distractions using heat and cold, visual and auditory distractions, memory tasks like counting, games, even humor can all be powerful pain management tools.[20,21,22] This gives us thousands of ways to distract the brain and dim pain.

"On the plus side, you've cured my back pain."

Credit: © Alex Gregory/CartoonStock

Caveat: It's Complicated

As always, some important caveats: Distraction isn't a panacea, and it's not a magical cure. When the distraction ends, as when medications wear off, pain is liable to return, especially if we aren't simultaneously targeting other ingredients in our pain recipe. This is also not a suggestion to stop attending to, or thinking about, your pain experience. It's actually extremely important to do these things. Moreover, distraction can be challenging: Brains have a frustrating tendency to fixate on the very things we try to push away. If I instruct you to not think about a pink elephant—its long pink trunk, its pink feet—that's all you're liable to think about.

Distracting from pain as a strategy for relief isn't the same as ignoring it, denying it, or pretending we don't have it. That said, allowing pain to dominate our cognitive real estate by constantly focusing on it and talking about it only gives it power, and amplifies the alarm.

New Directions and Next Steps

The human mind is endlessly fascinating, its ways mysterious and surprising. We can use our brains to stay warm when it's cold, like the monks meditating in the Himalayas. We can disrupt the Pain Cycle by breaking thought patterns that hijack our physiology. We can modify our mindset, examining which of our thoughts are true and which might be distorted—cultivating hope and reducing pain in the process. And we can harness the science of attention to identify our own invisible gorillas, and distract our brains from pain. One of the most successful ways to positively redirect attention is by engaging with the world—and most especially, the *people*—around us. That's one of the reasons why pain is social, too—the subject of the next chapter.

7

Pain Is Social

WHY FRIENDS ARE ACTUALLY MEDICINE

The Biology of Belonging

As I was growing up, my grandmother Marjorie was like a parent to me, and more: a confidant, pen pal, and ballast. When a brain tumor and subsequent series of strokes rendered her bedridden for the final years of her life, it became part of my weekly routine to visit and sit by her bedside. She lit up when she saw me, her energy visibly increasing.

On this particular day, I brought her a plant—at age ninety-four, she still loved greenery and all things that bloomed—and settled in to tell her stories. She shimmied up in bed, propping herself upright. Since the last stroke, she'd been steadily declining; her death, the neurologist said, was imminent. Instead of focusing on her emaciated frame, her hollow cheeks, or her trembling hands, I focused on her eyes. "Bright," I commented. "Cheerful. Full of life."

"Oh, honey!" she exclaimed. "You know I always feel better when you're here." The attending nurse nodded in agreement. "She eats more on the days you visit," she said. "She has so much more energy."

"Good medicine," said my grandmother, gently taking my hand.

* * *

Social medicine is fundamental to our well-being—so much so that it is quietly woven into the very fabric of our daily lives. It's that time you had the flu as a child, and your mother rubbed your back, fed you soup, and you *felt better*. It's when sharing terrible news with loved ones seems to halve the heartache. It's how community-based twelve-step programs support sobriety, and how having a gym buddy makes it more likely we'll actually exercise. It's the way our heart rate slows and our muscles relax when the cat curls in our lap and purrs. It's the way hope swells and spirits rise when a friend says "You're not alone."

And then, it's much, much more.

I first started studying pain neuroscience as an undergraduate at Brown University under the mentorship of a renowned pain neuroscientist. At the time, it baffled me that pain was biopsychosocial. That pain was biological was clear; that it was also emotional seemed intuitive. But social…? How did one's social life have anything to do with their chronic knee pain, their diabetes, their brain tumor? We just don't learn about this in school. Even in graduate-level training programs, it's exceedingly rare that any healthcare provider is required to take a course on social medicine. This includes nurses, physical therapists, physicians in medical school, and psychologists. The social domain of health thus remains among the most neglected in medicine.

But as I progressed in my studies of pain science, the following five pieces of evidence, laid out before me like the facts of a legal case, forever changed my perspective. The shift was gradual; it took decades of research and reading to untangle the surprising, powerful interconnections between social health and physical pain.

My hope is that it will take you only until the end of this chapter.

Exhibit 1. Social Isolation Changes the Brain

What's the worst punishment you can give a human being?

It isn't prison. When prison inmates are punished, they're tossed in "the hole"—locked away in solitary confinement, cut off from all contact with others. This practice has proven horrifically detrimental to human health, eroding the brain and increasing risk of premature death. It also triggers a host of mental health issues, including depression, anxiety, suicidality, and psychosis.

Just take Kiana. He was only seventeen when he was thrown in solitary confinement at one of America's most brutal prisons, Louisiana State Penitentiary. He was placed in a tiny cell, prohibited from seeing or talking to anyone else. There were no distractions, reading materials, or educational programs. Kiana languished there for sixteen months. Inmates referred to solitary confinement as "twenty-three and one," because prisoners spent twenty-three hours alone in their cells with only one hour to take a shower and take care of other business—if they were lucky. "The hardest part of living in solitary," he says, "is trying not to lose hope. You're stuck in your cell with just the voices in your own head and the cries of men who have already gone mad."[1]

Since his release from prison, Kiana has made it his life purpose to prevent others from enduring the same horrors. He leads campaigns for prison reform, and helps formerly incarcerated young men get job training and find purpose. But even twenty-plus years later, Kiana still has nightmares and flashbacks. It's no surprise that the United Nations considers extended solitary confinement a form of torture. What does it say about human beings that one of the worst things you can do to us is to isolate us from others…?

Humans are social animals. We evolved to be this way over many centuries for one fundamental reason: Social behavior helps us survive. Collaboration, cooperation, and communication help us secure food, water, and shelter. Community provides protection from predators, and assistance when we're vulnerable, sick, have babies, and grow old. Social behavior is thus written into our genetic code. In fact, it's so critical for our survival that our bodies evolved a mechanism to reward us for engaging in it. In the presence of others, our brains release neurotransmitters like dopamine, serotonin, and oxytocin that generate feelings of happiness, reward, connectedness, and pleasure. When we're social, our brains also release endorphins—our body's homemade painkillers. And when we're socially isolated? All of these chemicals crash.

There's a reason why loneliness hurts and rejection breaks our hearts. Indeed, a broken heart is often more painful, and more difficult to heal, than a broken leg. Social contact is so critical to our survival that isolation and loneliness can be experienced as physically painful—similar to other unmet needs, like thirst and extreme hunger. This is because "social pain," like emotional pain, is processed by the same regions of the brain as physical pain, mediated by overlapping neurochemicals.[2,3] It's no wonder, then, that opioids and other painkillers are so addictive: They blunt not just our back pain, but also the agony of loneliness, depression, and isolation.[4,5]

Exhibit 2. Loneliness Kills: Morbidity and Mortality

Just as social deprivation affects the brain, it also wreaks havoc on the body.

Two words we use regularly in medicine and healthcare are "morbidity," or the rate of disease in a population, and "mortality," or death. One little-known fact is that these terms are inextricably connected to our social health.

The evidence? Older adults who are socially isolated are more prone to illness, disease, and even death than those with social support.[6] There are multiple reasons for this. Social support means having others around to care for us in old age, take us to the doctor, and monitor our health and medications. But another, equally important reason may be a bit less obvious.

Social deprivation is a massive stressor on the human body—especially when it lasts. Chronic loneliness and isolation keep our stress system in a persistent state of activation, resulting in the overproduction of a stress hormone called cortisol. Chronic, high levels of cortisol suppress our immune system. This means that loneliness and isolation, especially over the long term, render us more likely to get sick if we are healthy, more likely to get sicker if we're already sick, and also more likely to die.[7]

The insidious, pervasive effects of loneliness aren't just reserved for the elderly. Former US Surgeon General Dr. Vivek Murthy released a groundbreaking report in 2023 exposing the impact of loneliness on *all* of us—specifically, on our physical health. His report, synthesizing years of medical research, confirms what we all suspected: the quantity and quality of our relationships affect not just our psychology, but also our biology. The health consequences of social deprivation include a 29 percent increased risk of heart disease, a 32 percent increased risk of stroke, a 50 percent increased risk of dementia, and a 60 percent increased risk of premature death.[8]

Not surprisingly, loneliness and isolation can also predict, trigger, lengthen, and amplify pain.[9] Data collected from over half a million people showed that social exclusion increased our likelihood of experiencing chronic pain. In fact, the lonelier we are, the more likely we are to report more pain in the future.[10] This reveals yet another cycle: The more we hurt, the more we tend to withdraw from people and activities. And the more we isolate, the lonelier we become. In turn, loneliness and isolation trigger and amplify pain.

Here's the underlying science: Social animals find safety in numbers. Over time, without the protection of our tribe, the brain enters a chronic state of threat detection to compensate for the lack of social support and the security it brings. A brain constantly on the lookout for threats is prone to magnifying input in case it indicates harm. *That sound, is it dangerous? What about that twinge in my leg? I'm all alone; do I need to find help?* Being in a constant state of high alert, combined with a body in a constant state of stress, turns the brain into a pain megaphone.

Chronic loneliness affects other systems in the body, too, contributing to systemic, chronic inflammation and disrupting our endocrine, metabolic, and immune systems. This physiological upset can even promote the development of diseases like cancer, heart disease, and stroke.[11]

But there's good news, too, because the opposite is also true: Social connection has tangible, quantifiable benefits. Being with others has the power to change our brains and bodies, improving our health, increasing longevity, and *reducing our pain*.[12] How? Social support activates changes in our immune system, bolstering our resistance to disease, protecting against illness, and speeding healing. It can decrease our cortisol levels and reduce inflammation. Spending time in community also alters our brain chemistry, boosting our mood and making us feel better both physically and emotionally. Having others around increases our motivation to engage in healthy behaviors that can, in turn, reduce our pain—exercising, for example, or reducing alcohol use. Think of how much more motivated you are to go to the gym when you have a friend to go with.

Ultimately, scientists have concluded that having a strong social support network is so powerfully linked to our physical health that it improves our odds of survival by 50 percent compared to those with a weak one.[13] Conversely, poor social health is as significant a risk factor for premature death and disease as smoking, obesity, physical inactivity, and alcohol consumption. It's true what they say: Friends really are medicine.

I see this in action every day as my clients demonstrate the many ways social medicine transforms pain. We witnessed this with Joyce, whose social prescription for back pain included joining water aerobics classes at her local gym and befriending the participants. We saw this with Fallon, who scheduled movie nights with friends as she healed from CRPS. And we saw this with Mateo, whose life and recovery were utterly transformed by the combined support of his parents, a dedicated treatment team, and Emmett the Explorer.

But despite the fact that social medicine affects all of my cases, no one has ever summarized its power as convincingly, and as beautifully, as Coach Murph.

Strength in Numbers

The cancer diagnosis came as a shock, Coach Murph, age sixty-six, told me. ("My dad is Mr. Murphy," he'd said. "Call me Murph.") The week before, he'd felt perfectly fine—living his life, raising his kids, and coaching his football team, seemingly healthy as a horse. The following week, after a routine doctor's exam, he was in the hospital receiving ominous news: He had cancer.

Coach Murph had always considered himself strong and resilient. But the diagnosis hit him like a semitruck. He hid it well on the car ride home, feigning optimism for his wife, Rose. She was his second shot at happiness, he told me—his second wife, and much younger. He didn't let himself cry until he was locked in the bathroom with the shower running. He didn't want his family to worry. His kids, identical twins, were only twelve.

Murph came to me to map out a plan for the pain before his treatment even began. "I'm a man who's always prepared," he said grimly. "That's why my football team always wins." After reviewing our biopsychosocial approach, Murph said he felt confident about the biological pillar of pain—he'd locked in a great team at the local cancer center. But the social piece, he said, felt more nebulous. Together, Murph, Rose, and I brainstormed forms of social medicine he could implement when the time came that might boost his mood, resilience, and resolve, and help ease his pain.

Murph's surgery was successful—but there were complications. After being sent home, Murph was in so much pain that he was quickly readmitted. Scans showed that the cancer was more widespread than they'd initially realized. Additional surgeries followed, along with radiation and chemotherapy. Murph spent months in bed. He lost his appetite, his hair, and his energy. It felt like the pain would never end. His mood bottomed out; resolve started to tank. Murph started thinking about his life—how full it had been, and how it would be okay if the cancer took him. When he expressed this to his wife, she became alarmed. Rose, who'd been lovingly tending to Murph on her own for months—in addition to single-parenting their two kids—started feeling the weight of hopelessness, too.

That's when Rose called me. "We need help," she said, her voice breaking.

Rose and I reviewed our social medicine plan and made some quick adjustments. That night, she sent out a fleet of emails to a vast network of friends, colleagues, football teammates, and family members. The response was instantaneous. From the following day onward, Murph had a near-daily visitor. He later told me that the impact this had on him was palpable. He still felt awful, but his energy and mood slowly started to improve—along with his resilience. He looked forward to his guests every day; they were bright spots in the darkness, bringing with them strength and energy he could borrow until his returned. For Rose, visitors served as much-needed sources of respite, giving her time to take care of her kids—and herself.

Cards and flowers started arriving. Soon, Murph's room was full of blooms. Every time he looked at them, they buoyed his spirits. Friends at church wrote to say they were praying for him, and his priest offered a healing prayer during Mass. Members of the football team, guys he'd coached for years, called with well-wishes. They even asked for advice, requesting analyses of videos of the opposing team to give them an advantage in the next game. "It made me feel useful," he later told me. "Those small tasks gave my time at home some meaning."

The social medicine was so nourishing that Murph decided to take it with him to the hospital when it came time for his next surgery. He brought the blanket his sister had knit for him, and slept under it every night. His twins smuggled his football jersey with "Coach" emblazoned on the back into the hospital. It made him feel stronger. Friends and relatives dropped off meals. Whether he could eat or not that day didn't matter—the food was a form of love. His friends weren't in the room, but he could feel their presence.

Murph's oncologist, a warm, quirky doctor who wore colorful bow ties, placed his hand reassuringly on Murph's arm. "You're not alone," he said. It gave him inexplicable strength. As part of his recovery, Murph's medical team urged him to walk. He moved slowly and needed help, but the walks offered unexpected social medicine, too. Passing other patients in the hallways, Murph felt an immediate kinship. They eventually started talking. Grace, a young history teacher, was battling her third round of cancer. She'd beaten them all, and she'd beat this one, too, she said fiercely, her eyes blazing. Her bravery was contagious. Grace told him about an online cancer support group, and Murph joined. Having a community that understood his journey—and had survived it—was invaluable.

I hadn't heard from Murph for a few months when he emailed to request an appointment. He was finally out of the hospital and on the road to recovery. I asked about his progress, and how the social medicine had affected him. He didn't hesitate. "I mattered," Murph said quietly. "I could see it, feel it, and taste it. People were out there thinking about me, sending their love, and praying for me. Knowing that my community was holding me gave me strength and the will to survive."

Exhibit 3. Pain Is Contagious: Why You Feel My Pain

We are so fundamentally interconnected, our brains so intrinsically wired to be social, that we can feel *each other's* pain.

In a room full of babies, as any parent will attest, if one baby cries, many are likely to start crying. This "emotion contagion" is also the reason we imitate the posture and facial expression of the person we're talking to; why we sob during sad movies, even though the sadness is not our own; and why laughter is contagious.

But it isn't just emotions that are infectious. Pain is contagious, too.

In fact, the more connected to someone we feel, the more we feel their pain.[14] A look inside the brain illuminates why this happens. When we witness others' pain, the same regions of our brain are activated as when we experience pain ourselves.[15,16] These regions include the insula, prefrontal cortex, and anterior cingulate cortex, parts of the brain described on page 35 known for their emotion-pain overlap. This is why we cringe and look away during a boxing match when the blood becomes too much to bear, why we feel a sympathetic twinge while listening to stories about gruesome injuries like Mateo's, and why parents inevitably suffer when their children do. In neuroscience, this phenomenon is known as "empathic pain," or the social transfer of pain. It is mediated by mirror neurons in our brain that help us mirror, or relate to, the pain of others in our tribe.

And this makes evolutionary sense: Understanding, empathizing with, and reacting to other's pain is adaptive. It helps us forge and strengthen bonds with members of our community, a social glue ensuring that we not only give help, but also reciprocally receive help when we need it. This social superpower is so important that other social animals—dolphins, monkeys, mice, crows—engage in it, too. At its extreme, our brain's empathic powers even contribute to mysterious conditions like Couvade syndrome, also known as a "sympathetic pregnancy." In this condition, a nonpregnant, biologically male partner develops symptoms of pregnancy, which can include morning sickness, breast tissue tenderness, hormone changes, weight gain, food cravings, even labor pains. While it may sound unusual, this syndrome is more common than we might think: Up to 72 percent of expectant fathers worldwide experience at least one symptom during their partner's pregnancy. The symptoms typically resolve once the child is born. Rather than pathologizing this condition, we might see it for what it is: an exquisite bid by our beautiful brains to empathize with our most important tribe member during a critically important time of life.

Exhibit 4. Toxic Relationships Can Make Us Sick

Just as healthy relationships help us flourish and heal, toxic, unhealthy relationships can do the opposite.

Abusive, traumatic relationships are notorious pain triggers and amplifiers. While physical abuse is often overt, leaving behind bumps and bruises, other kinds of abuse, like verbal or emotional, may be more subtle. And sexual abuse might be hidden from the world entirely. But these forms are no less toxic. One particularly noxious form of abuse is the silent treatment, a tool used to exclude, ostracize, and force social disconnection. Known as a "social death," its impact on human health is insidious.

Being alienated and shunned, whether from our family of origin, religious group, friend group, or another valued community, can trigger and exacerbate depression, anxiety, suicidality—and pain.[17,18]

Dr. Kipling Williams, for example, recalls the time that he was reading peacefully in a park when a Frisbee rolled onto his blanket. The two men who'd been tossing it, complete strangers, invited him to play. After a few rounds, the men began to exclude Dr. Williams, throwing the Frisbee only to each other. Despite the fact that he was a grown man—a successful doctor with friends of his own—being excluded "hurt," he says. It turns out that social exclusion is a form of punishment *designed* to hurt. Dr. Williams was eager to explore it further.

An assistant professor of psychology, Dr. Williams decided to study ostracism in his lab. He and his team created experimental scenarios in which an individual was ostracized to observe how it affected their brains and bodies. In one experiment they called the Scarlet Letter Study, a red letter "O" was affixed above a team member's office door.[19] The group was instructed to completely shun the marked member for the day. Forms of exclusion included not talking to, smiling at, or even making eye contact with the target. The team conducted this experiment five times with five different people, each of whom recorded their thoughts, feelings, and reactions. Without fail, the shunned target experienced varying degrees of physical and emotional pain—including discomfort, insomnia, anxiety, and loss of self-esteem.

Years of data indicate that even brief periods of shunning and estrangement can trigger stress and despair, rupture our sense of identity, and erode our sense of control. Longer-term estrangements can trigger suicidality, trauma symptoms, physical signs of chronic stress, and pain. Over the last decade, Williams has become one of the premier leaders in this niche field. "No matter how people are left out," he writes, "their response is swift and powerful, inducing a social agony that the brain registers as physical pain."[20]

There's a reason for this, one rooted in our very neurobiology. Social exclusion actually increases the sensitivity of our pain system, a phenomenon known as "ostracism-related hyperalgesia." Humans can experience this on an individual level—for example, a mother estranged from her children, or a lonely child ostracized from his schoolmates—and we can experience it on a larger scale, too. Racism, in which an entire race is stigmatized and marginalized, is also a well-established pain amplifier.[21]

Scientists hypothesize that, similar to physical pain, social pain may serve as a danger alarm: a signaling system alerting us that critical social bonds have been damaged, urging us to either repair them or forge new ones.[22] Given that social relationships are as fundamental to our survival as food and water, this makes sense, as it is as much to our detriment to ignore social breaks as it is to ignore a broken jaw. And as with the physical discomfort of damaged bones and body parts, the discomfort of social disconnect may similarly compel us to take action and make repairs.

Of course, relationships don't need to be abusive to be unhealthy. Toxic relationships come in myriad other forms. Relationships characterized by dishonesty and gaslighting, those that are chronically hostile and stressful, and acrimonious splits are also destructive. These kinds of interpersonal interactions can trigger a sustained sympathetic stress response, hijack our immune system, and crank up the Pain Dial. Breakups and interpersonal conflicts, for example, have well-known physical consequences—triggering muscle tension, headaches, and stomachaches, and disrupting sleep and appetite. In some cases, toxic relationships can even increase our risk of developing illness and disease.[23]

If toxic relationships can make us sick, and if supportive relationships can reduce pain, there must be ingredients absent from our recipes—major medicines missing from our drug cabinets. Let's take a closer look at one of these.

Exhibit 5. The Healing Power of Touch

Picture this: Two tiny, premature babies are kept alive in a neonatal intensive care unit (NICU) incubator. When sufficiently stable, they are moved to a transitional unit for additional care and monitoring. The babies drink the same amount of formula and consume the same number of calories. One baby, however, receives an additional fifteen minutes of touch three times a day for ten days. The baby who receives this additional touch gains 47 percent more weight per day than the baby who did not.[24] She also spends more time awake and active, exhibits more developed motor skills, and is able to leave the hospital a full week sooner. Surprisingly, these gains are retained over time: When the babies are again assessed eight and then twelve months later, the infant who received additional touch is still in a higher weight group, and again scores higher on mental and motor tests. Except there is one caveat to this true story.

It isn't a story of two babies.

It's a story of billions.

Including you, and me.

This is because touch is fundamental to being human. We are quite literally created inside another body, enveloped by her warm womb. Touch is therefore the very first sense we develop in utero, before even sight or hearing. Touch isn't just pleasant and calming; it's necessary for our very survival. Studies of orphaned babies and infants indicate that touch deprivation in early life can result in stunted growth, deficient motor and cognitive development, a weakened immune system—and even death. Those who have experienced touch deprivation describe it as a physical ache: a thirst, only in the skin, and for human contact rather than water.

Social touch—being touched by another—has been shown to confer marked health benefits, reducing levels of stress hormones and boosting our immunity. As we saw with the preemies in the NICU, touch encourages growth and stimulates brain development. Skin-to-skin contact with a baby after birth produces measurable improvements in her heartbeat, weight, sleep, energy, resistance to infection, ability to learn, and overall resilience, which is why doctors immediately place newborns upon their mother's chest. Touch also prompts the release of oxytocin—a feel-good neurochemical that promotes bonding, attachment, and closeness. Oxytocin is released when we hug a loved one, get a back rub, or even hold hands.

Touch is medicinal in other ways, too. It has been shown to help patients recover from burns, reduce symptoms of eating disorders, and boost immune function, giving individuals with HIV better ability to fight the virus. Touch can help us in the workplace, too: Employees who received massages showed a meaningful reduction in blood pressure, anxiety, and stress, and demonstrated improved speed and accuracy, compared to those who didn't.[25] Touch is also a legendary analgesic, or pain reliever.

When you painfully crack your knee on the corner of the kitchen table, what's the first thing you do? You rub it. There's a reason we do this, and it is because *touch reduces pain*. The decrease in pain triggered by touch is called "touch-induced analgesia." It helps explain why rubbing a stomachache makes it feel better, why a hug can feel medicinal, and why kissing a baby's boo-boo makes him stop crying.

Neuroscientists hypothesize that touch exerts these analgesic effects by interrupting pain processing. For starters, touch messages from the skin rush to the spinal cord, where they inhibit the transmission of danger messages from the damaged body part to the brain. Because fewer danger messages reach the brain, the pain alarm quiets. So, while the injury remains the same—our damaged knee still swells—touch helps us hurt less.

Touch, particularly being touched by trusted others, also reduces activation in regions of the brain that process pain, soothing our nervous system and quieting our fight-or-flight response.[26] It simultaneously triggers changes in our body: Our muscles relax, blood pressure lowers, and heartbeat slows. Our breathing becomes more regular and less erratic, and levels of stress hormones drop. Feelings of safety increase, and our immune system gets a good boost. Touch also stimulates the release of endorphins, our endogenous opioids, which play a fundamental role not just in pain relief, but also in social bonding and comfort in both non-human primates and humans.[27,28]

The Verdict

An old proverb advises us that sharing our joy doubles it, while sharing our pain cuts it in half.

It turns out that this is actually true.

The evidence is in, and the verdict is clear: Pain treatment comes in many forms, and social medicine is one of them. Strong connections and communities lead to a happier—and healthier—life. This is the reason my grandmother, and yours, feels better on the days we visit. True healing includes bolstering community and social connection; setting boundaries around toxic or abusive relationships; utilizing touch and safe sensations, like heat and cold; and tending to the pain of splits, separations, and deaths. This widely available medicine comes in many forms, including supportive interactions with peers, providers, community groups, family, and even safe strangers online. What makes social medicine distinct is that *we* have the power to prescribe, dose, and inject ourselves with as much or as little as we choose.

And with that, Your Honor, I rest my case.

8

Pain Is Environmental

THE SEED IS ONLY AS GOOD AS ITS SOIL

A Flamboyance of Flamingos

Flamingos, widely beloved for their elegance and splendor, grow to be up to five feet tall, drink with their heads upside down, and—a relative rarity among animals—are bright pink. Their name is derived from a Spanish or Portuguese word meaning "flame-colored." Together, a group of them is known as a "flamboyance." One might reasonably assume that flamingos are simply born this way, with feathers aflame.

But one would assume wrong.

At birth, newborn flamingos emerge from their eggs a dull, slate gray. Their hallmark color, despite being fused with their identity, is not inherited. It is not genetic; it is not encoded in their DNA. Rather, a flamingo's pinkness is completely and utterly contingent upon its environment.

Beta carotene is a colorful pigment that turns carrots orange and ripe tomatoes red. It is also found in the shrimp, algae, and brine fly larvae that make up the bulk of a flamingo diet. As the birds consume these foods, enzymes in their bodies metabolize the pigment and distribute it throughout their bodies—turning their feathers, and even their skin, a rich, electric pink. A flamingo raised in an environment lacking these food sources emerges from its egg dull, drab, and gray...and it stays this way.

We witness a similar phenomenon with things living closer to home. When we fail to water our gardens or provide them with sufficient sunlight, our plants wither, turn brown, and die. But give them rich, nutritious soil and offer a sunny, well-hydrated plot, and there they will bloom, flourish, and thrive. In fact, when we "fix" a plant, we really only fix the environment in which it grows.

As with birds and botany, humans' environments matter, too. The genes we inherit from our biological parents give us our genotype: our DNA and genetic blueprint. And when these genes interact with our environment, this yields our phenotype: all of the observable characteristics of the human, plant, or flamingo you see before you. All of them, from our height to our health. These influential environmental factors—which include culture, race, and ethnicity; sex and gender; socioeconomic status and access to care; and environmental stressors like trauma and abuse—are called social determinants of health, and they impact every single thing about us. EVERYTHING. In fact, many things we presume are a simple matter of biology—depression, for example, or cancer, and certainly pain—are *as much environmental* as they are biological.

Because this topic could fill multiple books (and it has), I will focus here on three of these environmental determinants of health:

1. Culture, race, and ethnicity
2. Sex and gender
3. Trauma

My hope is that this chapter will help you better understand the soil *you* grow in: the environmental and sociological ingredients that contribute to your pain recipe.

Culture, Caregivers, and Pain Gods

We tend to assume that humans across the globe all understand pain the same way because of its universal evolutionary purpose. But while all humans likely process pain similarly at the neurobiological level, if not nearly identically, our understanding of pain—what it is, what it means, how to respond to it—is tethered to our culture.

From a young age, we learn what pain means from our caregivers and our community, the things that are expected to hurt and not hurt, the ways in which we're expected to express our pain, and how others are likely to respond to us. This is in part due to a phenomenon called "social modeling," and it affects us from the day we are born.

"Social modeling" means that the people in our environment—family members, coworkers, religious leaders, actors on TV, community members, even the social media accounts we follow—all serve as role models and teachers, exemplifying what pain is, the reactions it should elicit from others, and how we're expected to behave in response to it.[1]

This is why, after a child falls on the playground, the first thing she's likely to do is look up at her parent's face. If her parent appears panicked and cries out with fear and alarm, the child will likely cry. But if her parent responds calmly and reassuringly, and helps the child distract and resume activity, she is *less* likely to cry, and more likely to resume play. In this way, the people around us unwittingly, and constantly, shape our pain experience.

Of course, our culture is much bigger than our caregivers. Every civilization and religious group in existence has its own concept of pain. Some cultures, for example, believe pain is a test of faith; others believe that pain is a punishment for past sins. Some cultures worship gods of pain: the ancient Greeks had the Algea, three goddesses who embodied physical and emotional pain, also known as the Dolores in Roman (from the Latin word *dolor*, which translates to pain, sorrow, and grief). And the Finnish have Kiputyttö, a spirit believed to ease pain by taking it to places it can't be felt, like the bottom of the ocean. Buddhists believe pain is an intrinsic part of being born in a human body and an integral part of life—something to accept, embrace, and grow from rather than fight or fix.

Every cultural group also has its own unique language of distress and suffering, tied to whether they value emotional expression, or stoic repression. More emotive cultures might model and expect a vocal display, like shouting and cursing, while cultures that value stoicism tend to encourage hiding pain and carrying on rather than making it public.

Our pain response is also influenced by what we've learned about pain's significance: Does our culture consider our symptoms "normal," or "abnormal"? Researchers examining cultural responses to pain in Australia and Nepal found that Australian Aborigines and rural Nepalese don't seek medical services for their chronic back pain even when offered, despite its ubiquity—one-third of men and half of women reported back pain when asked.[2] Why...? Because in these cultures, back pain isn't perceived as a "medical issue" that needs to be fixed—but rather a natural, normal part of aging.

Consider this: In a small village in Jordan, a group of devout Sufi men gather in a spartan room to pray. There is music, chanting, and singing. The men sway rhythmically, eyes closed; some practice meditation and breathwork. The leader of the ceremony unwraps a leather pouch, revealing knives that will be used for a religious ritual demonstrating faith in Allah and a belief in His ability to heal. A few of the men then use these blades to pierce their skin. Instead of fear or dread, there is eagerness to participate; and despite the puncture wounds, no one cries out in pain. Afterward, there is a sense of deep peace and communion. For Sufis, this kind of pain isn't suffering—it is transcendental, a way to rise above the physical body and get closer to God.

On a more granular level, our "culture" also refers to the communities and subcultures with which we choose to identify and affiliate. The marathon-runner community, the Kung Fu community, and the BDSM community (which consensually engages in sexual practices involving bondage, dominance, submission, and masochism) to which we belong are cultural influences, too. Each conceives of pain, and how to respond to it, in its own unique way, which inevitably shapes our understanding of it. It may be normal and accepted to pursue pain on purpose, for example, to inflict pain on willing others, or to enjoy pain in these subgroups—regardless of whether this differs from our broader cultural norms.

Deeper Than Skin-Deep: Race, Ethnicity, and Pain

Pain is also influenced by groupings over which we have much less control, like our race, ethnicity, and socioeconomic status. These impact not only our pain experience, but also others' perceptions of our pain—that is, how others *think* we feel. Serena Williams, tennis star and champion, can speak to this firsthand.

The day after giving birth, Serena lost feeling in her legs and began having trouble breathing. She suspected she was experiencing a pulmonary embolism—a condition she'd been diagnosed with years prior in which life-threatening blood clots obstruct blood flow to the lungs. She immediately requested a test and blood-thinning medications. Instead, the nurse rejected Serena's request, suggesting that she was simply experiencing "confusion." When Serena's doctor finally agreed to the tests, scans revealed she'd been right all along: Blood clots had indeed formed, and Serena required immediate surgery. She has since wondered what would've happened to her had she not been a famous athlete. "In the U.S., Black women are nearly three times more likely to die during or after childbirth than their white counterparts," says Serena. "Being heard and appropriately treated was the difference between life or death for me."[3]

Humans tend to make assumptions about the pain of other races, ethnicities, and cultures, and these are often based on ingrained, and notoriously toxic, stereotypes. One false belief about Black Americans is that Black people have "thicker skin" than white people, and thus have a higher pain tolerance.[4] It isn't true. But this stereotype, which dates back to the slave trade, contributes to racism in medical settings, disparities in health care, and discriminatory prescribing practices. Even now, Black Americans are systematically prescribed fewer pain medications than their Caucasian counterparts.[5]

Moreover, Black patients aren't always believed when they report their pain. This doesn't just happen to women like Serena. The pain of Black men and even Black children is also systematically dismissed more than white people's pain. The *Journal of the National Medical Association* reports that Black patients of both sexes are twice as likely to have their pain underestimated compared to other races and ethnicities.[6] In another study of nearly a million children with appendicitis, Black children were five times less likely to receive opioids for severe acute pain than white children.[7]

Our beliefs about wealth shape our perceptions of pain, too. Across *all* races and ethnicities, individuals of low socioeconomic status are more likely to be perceived as "exaggerating" their pain than wealthier individuals. Research suggests this perception is a dangerously false one: A massive study of more than nineteen thousand people conducted over twelve years found that the economically disadvantaged experience more pain than wealthier individuals, and also experience more severe pain.[8]

Science offers some hypotheses for these findings, and not a single one has to do with inferior genes. For one, it's well established that ethnic minorities and the economically disadvantaged experience a disproportionate burden of illness and disease, largely due to environmental factors. These include access to affordable, quality healthcare; difficulty obtaining insurance; inadequate access to affordable, healthy foods; and unsafe living conditions.[9] Second, a growing body of research shows that racism, marginalization, and stigmatization, all massive stressors, sensitize the pain system and increase inflammation, leading to more pain at lower thresholds.[10,11]

The result? A growing disparity in the incidence, prevalence, and treatment of pain—predicted not by genetics, disease, or injury, but by race, education, and socioeconomics. These ingredients have all been hidden in your pain recipe, unannounced and unaddressed.

And there's yet another social determinant of health we haven't explored—one that applies to each and every one of us: our sex.

Take It Like a Man: Sex, Gender, and Pain

If you think the experience of pain has nothing to do with sex or gender, think again.

When you hurt, do you "cry like a girl"? Or have you ever been instructed to "take it like a man" after an injury?

Distinctions between the sexes are glaring, starting with this one: The clear majority of people who experience chronic pain are women.[i,12] Women are at higher risk for developing chronic pain, and report more severe pain, than do men.[13] "Whether or not women report pain more or have pain more," says Dr. Carolyn Mazure of the Yale School of Medicine, who studies sex and gender differences in medicine, "I think we could say that both may be true."[14]

However, we still aren't sure, because important data about women's pain is only just emerging. Why, you ask…? Eighty percent of pain studies have been conducted only on males, primarily human men and male rodents. This disturbing trend is universal across medical research, and has been ongoing for decades. In fact, the vast majority of scientific studies have historically excluded females—collecting data *only* from males, and generalizing it to everyone else.[15] There is even less data on the diverse spectrum of other gender identities. Despite this, it has become abundantly clear that gender is a formative and critical ingredient in our pain recipe.

As with all things to do with pain, these differences aren't a simple matter of biology. Rather, sex- and gender-based pain differences are biopsychosocial: the by-product of intersecting biological, psychological, and sociological factors that affect not only how we experience pain, but also how we rate it.[16]

---

i As defined by the World Health Organization, *sex* here refers to biological, heritable attributes such as genes, sex hormones, and reproductive anatomy. *Gender* refers to socially constructed roles, behaviors, expressions, and identities. Source: Kaufman, M., Eschliman, E., & Karver, T. (2023). Differentiating sex and gender in health research to achieve gender equity. *Bulletin of the World Health Organization, 101*(10), 666.

In American culture, for example, men are traditionally encouraged to suppress their emotions. If men do express pain, they're at risk of being labeled "weaklings" or "sissies." Women, on the other hand, typically acculturated to show emotion, tend to express and report pain more freely than do men. However, when women do, they're much more likely to be accused of exaggerating, faking, being overly emotional—or, worse, dismissed as "crazy" or "hysterical." This has been ongoing for centuries. Hippocrates, the first to use the term "hysteria," decided it was a disease of the uterus—despite the absence of a single shred of scientific evidence to back this claim. Freud later parroted him, labeling hysteria "exclusively female." The impact of this pervasive misogyny has trickled down through generations, and with catastrophic results.

Here's one: Women are twice as likely to die of cardiovascular events than men.[17] One of the reasons for this is that women who present to the ER with chest pain are twice as likely to be diagnosed with "mental illness" and sent home compared to men with the exact same symptoms. This is not an exception—it is the rule. Studies have found that healthcare providers are much more likely to dismiss women's pain than men's, attributing it to psychological causes like anxiety, stress, and other modern versions of hysteria. This tendency to attribute women's pain to mental health issues and emotions alone means that women's *actual* diagnoses are often missed entirely.

Ann, age sixty, recently shared her story with the *Washington Post*. "I started to experience strange sensations along my face, extending along the right side of my body," she says.[18] But when she told her primary care team, they scoffed, she says. Her symptoms worsened; she developed double vision and blurriness. When she asked her providers if it could be a brain tumor, they said no, and sent her home. Six months later, she says, "I lost consciousness and had an MRI." It was a brain tumor.

Gender biases affect our treatments, too. Sometimes they delay and prevent appropriate care. Women are significantly less likely than men to receive appropriate pain medications in the emergency room and after surgery, and are more likely to instead be prescribed sedatives to "calm down."¹⁹ "There's a pain gap, but there's also a credibility gap," says Anushay Hossain, author of *The Pain Gap: How Sexism and Racism in Healthcare Kill Women*. "Women are not believed about their bodies—period."

While these heteronormative stereotypes aren't always true, it's clear that our pain ratings—as well as our pain treatments—are just as reflective of culture and gender as they are of injury or illness.

Trauma and Pain: The Body Keeps the Score

Eleven Trigger Points, Four Bullets, One Gun

For Hallie, the headaches arrived first, when she was just fourteen. Hallie's neurologist had told her parents that she would likely grow out of them after puberty. Instead, they'd only gotten worse. As she aged, the pain spread to her neck and shoulders, then down into her hands and the joints of her fingers. And as pain worsened, so did fatigue, brain fog, and insomnia. Then came the nausea and stomachaches. By the time she was twenty, Hallie was only able to participate in life and activities when pain permitted. Other days, the pain was so intense, and her energy so low, that the mere act of washing dishes was all she could accomplish before climbing back into bed.

Batteries of tests and scans followed her into adulthood: blood tests, X-rays, CT scans, MRIs, nerve conduction tests, stool samples, physical exams. When all tests came back negative, Hallie was told it was likely "anxiety and stress." But despite her best attempts at stress management, Hallie's pain continued. Finally, when Hallie was twenty-nine, a rheumatologist diagnosed her with fibromyalgia. He explained the diagnostic criteria. Previously, he said, fibromyalgia was diagnosed only if a patient had pain in eighteen or more "trigger points" located across the body. Then, that number was reduced to eleven. Now, he said, experts had carved the body into five regions, and she only had to have chronic pain in four of them to earn a diagnosis.

Hallie's eyes widened with surprise. This made no sense to her. As everyone with fibromyalgia knows, on some days, eighteen body parts might hurt; on other days, only two. And on other days, the pain might spread to a new body part that had never hurt before. Pain severity constantly waxed and waned, too. How could one base a diagnosis on the location of pain, or number of body parts, when those locations and numbers constantly changed? But she was no expert, and the diagnosis seemed to fit. And so, it stuck. Her research suggested the outlook was grim: The internet informed her that fibromyalgia "had no cure." Did that mean her pain was permanent?

Friends and family struggled to understand her diagnosis. For most of her adulthood, Hallie had managed to work, see friends, exercise, date, even take photography classes. She *looked* fine, so people assumed she *felt* fine. But she didn't feel fine. Her illness was invisible, one that came without a wheelchair, bandages, or blood. And as the pain spread, it took a massive toll not just socially and physically, but also financially. Her insurance company had repeatedly denied coverage for any treatment that wasn't a drug, forcing her to eat into her savings. Pursuing her photography passion quickly became impossible, even though being a photographer was her lifelong career goal. She was lucky to cover her rent. Finally, when she could no longer afford to live on her own, she moved back in with her parents.

By the end of her second week at home, she realized not much had changed from the chaotic childhood she remembered. Her parents still got into ugly fights that lasted through the night. She often woke to the sounds of shouting and broken glass. Hallie's health, rather than improving, declined. She spent weeks in bed, headphones strapped to her ears, stubbornly ignoring the chaos around her. But the longer she lived at home, the worse her pain got. It migrated into her lower back, then around into her pelvis. Urination was painful, sex out of the question. She stopped dating. The pelvic pain went on for so long that she started wondering if she'd ever be able to have children. This was devastating for Hallie; she'd always wanted to be a mom. Changing her environment had indeed changed her pain—but for the worse. It didn't make sense. Why would living at home make her body hurt more?

Hallie was the lowest she'd ever been when her favorite doctor, an ob/

gyn she'd been seeing since her teens, forwarded her a podcast episode in which I described the relationship between trauma and pain. When she heard it, Hallie sank to the floor and wept. Why had no one ever explained this trauma–pain connection before? Things suddenly made sense. If only she'd known, she told me later, the information would've changed her life. She would've made the connection and told all of her providers.

She would've told them that she'd dreaded nighttime her entire childhood, a time when her father habitually sat in the den and drank. He was a violent drunk, prone to fits of rage. One night when Hallie was fourteen, her father began smashing glass vases against the wall. Next came water glasses, a computer, and every other fragile, breakable object he could find. Shards of glass exploded in the air and skittered across the floor. When Hallie's mother tried to stop him, he hit her. Hard.

Afraid for their safety, Hallie pulled her little brother, then age eight, into the bedroom and secreted him out the window. She prayed her father was too preoccupied to notice. They stayed on the roof the entire night despite the cold, curled together like kittens. When the sun rose, and the house was still and silent, they clambered back inside, pulled on their school clothes, and walked to the school bus as if everything was normal.

To them, it was.

Hallie's mother filed for divorce just a few weeks later. That night, her father came home drunk. Hallie could always tell by his heavy footfall and the way he slurred his words. In his hand, he held a small, black handgun that glinted in the hallway light. The gun, he said, his eyes hard and ugly, had four bullets: one for each of them. Hallie, her stomach in her throat, pulled herself up the full length of her fourteen years and attempted to calm him using her bravest, most soothing voice. Her mother finally enticed him into the backyard with a six-pack of cold beer—but not before he'd punched a hole in the wall the size of a watermelon. Hallie, barely breathing, head pounding, called 911.

Just as her parents' volatile dynamic hadn't ever recovered, neither had Hallie's body. More than a decade later, Hallie still struggled to feel safe, especially at night. It was as if her brain was waiting for the Next Bad

Thing to happen. She was often edgy, on high alert, and oddly attuned to sounds: Any loud noise, even something as innocuous as a car door slamming, made her jump out of her skin.

When she heard the podcast and finally made the link between trauma and pain, Hallie felt relieved. She'd always sworn her pain had no pattern, that it was unrelated to her family history. Why would it be? But this new orientation to suffering helped things click into place: her sleep issues, hypervigilance, and fibromyalgia. Looking back, it seemed clear that everything had always been connected. It was so logical—and yet so astonishing. For the next few weeks, the phrase that danced inside her head was "the brain is connected to the body 100 percent of the time."

Hallie came to find me, and her pain recipe poured out. Trauma was at the top of her list. As part of her treatment, she decided to limit her exposure to toxic, triggering relationships, which meant spending less time at home. It took some doing. With the help of a friend, she eventually moved out of her parents' house and into her friend's guest room. Setting these boundaries seemed to help. With assistance from her mom, Hallie was able to afford psychotherapy and biofeedback, which taught her how to manage and mitigate physical trauma responses. She learned relaxation and self-soothing strategies that helped her regulate her heart rate, muscle tension, and startle responses. She learned how to turn down her brain's alarm messages, the ones constantly telling her that she was in danger even when she wasn't. Recognizing her body's desperate need for soothing, Hallie also signed up for outdoor yoga, which gave her a dose of sunshine and got her moving again. Sometimes, it hurt—but her new awareness that fibromyalgia pain didn't mean danger was a complete game-changer.

Hallie also changed her diet, cutting out caffeine and alcohol. A longtime coffee fiend, it proved more difficult than she expected. But there was no doubt it improved her sleep. Before she went to bed every night, Hallie listened to guided meditations that transported her from the guest room into nature—grassy fields and ocean views. It relaxed her so completely that her muscles felt like they were melting. Finally, Hallie learned sleep hygiene skills that helped her ditch the sleeping pills she'd been taking for years.

It was challenging work. Progress was gradual but sure. As the fatigue and lethargy began to wane, Hallie's motivation and energy increased. Her gastrointestinal issues seemed to respond well to the dietary changes, relaxation, biofeedback, and meditation. Her head, neck, and arm pain slowly started to fade to background noise. Back and pelvic pain were more persistent, but became less incapacitating. Once she was able to type again, she reapplied for her old job. Her boss, rooting for her from the beginning, was thrilled to have her back. Each small win motivated Hallie to keep chipping away at her pain recipe, frustrating as it was.

She was eventually able to resume her beloved photography classes, which brought her great pleasure and joy. In time, the pelvic pain that had once canceled all thoughts of having children finally started changing, too. Hallie even allowed herself to imagine that she might actually have a family one day.

Hallie continued to have occasional flares—sometimes predictable, like when she went home to visit her parents, and sometimes unexpected. Except now they didn't scare her, and they weren't debilitating. Instead, when the pain came, she simply implemented her tools. They helped immensely, and she had *lots* of them.

Last month, Hallie moved in with a man she met in her photography class.

They are trying for a baby.

A New Understanding of Trauma

Of all the social determinants of health, the one that has gotten the most attention in recent years is trauma. This buzzword is now so overused in pop culture that it seems to have lost its meaning.[ii] But while the term may have been watered down, its impact on the human body has not.

Trauma and chronic pain are best friends, co-occurring up to 80

---

ii Trauma as defined here is from the *Diagnostic and Statistical Manual of Mental Disorders, 5th Edition, Text Revision (DSM-5-TR)*: "exposure to actual or threatened death, serious injury, or sexual violence," either as a victim or a witness.

percent of the time: people with chronic pain frequently have a history of trauma, and people who have experienced trauma frequently develop chronic pain.[20,21] In fact, the presence of one is known to increase the likelihood—and the severity—of the other. This is no coincidence. Thanks to Dr. Vincent Felitti and colleagues, of this much, we are certain.

In 1998, Dr. Felitti was treating obesity at Kaiser Permanente hospital when he made a puzzling discovery. The primary driver of his patients' weight didn't seem to be food. Beneath the surface, it was about something much more sinister.

Despite wanting to be healthier, many patients were ambivalent about losing weight. Some even referred to their heft as "protective." It didn't make sense. From what did they need protection?

When Dr. Felitti interviewed his patients to investigate, a surprising pattern emerged: A disproportionate number had survived childhood abuse. And this abuse seemed to be linked to their obesity. One twenty-three-year-old, a woman who'd packed on 150 pounds after being raped, explained the connection: "Overweight is overlooked, and that's the way I need to be."[22] For her, the extra weight was protective—a trauma response that offered a sense of safety from sexual predation. Her obesity wasn't the problem; it was the *solution*.

If there was a hidden relationship between trauma and this disease, the scientists wondered, could trauma also play a role in others? Felitti and his team decided to find out. Pairing up with the US Centers for Disease Control and Prevention, they launched an investigation into the ways in which childhood traumas—otherwise known as adverse childhood experiences, or ACEs—affect human health. They surveyed nearly twenty thousand people, inquiring about physical, sexual, and emotional abuse; neglect; domestic violence; parental mental illness; and other early adverse experiences. They simultaneously assessed all manner of physical and emotional health outcomes. Their findings shocked the medical world, and continue to reverberate today.

Their groundbreaking work, known as the ACE Study, revealed that childhood traumas were common, destructive, and among the most

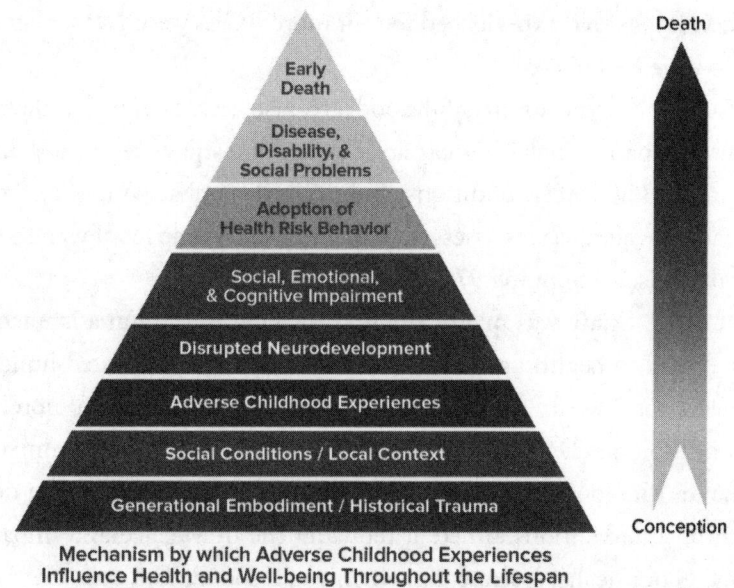

Trauma predicts disease, disability, and chronic pain.
Credit: ACE pyramid courtesy of the US Centers for Disease Control and Prevention

important determinants of physical health ever uncovered. Exposure to ACEs significantly increased risk of developing diseases like obesity, cancer, depression, migraine, and lung disease.[23,24] And as the number of childhood traumas increased, so did the physical and mental health issues. Patients who'd endured multiple childhood traumas, for example, were 460 percent more likely to be depressed, and had a 1,220 percent increased likelihood of having attempted suicide, than patients who'd experienced none.

The same pattern has been found with chronic pain. The more traumatic events we experience, the greater our risk of developing chronic pain and disability.[25,26] Among adults who have experienced adverse childhood experiences, the incidence of chronic pain is double that of those who've survived none. But pain after a trauma doesn't always wait until we grow up. In a study of nearly fifty thousand children, those with one ACE were 60 percent more likely to have chronic pain compared to those with none.

And children who'd experienced four or more ACEs were *170 percent more likely* to have chronic pain.

Traumas that occur in adulthood are also risk factors for developing chronic pain—including car accidents, domestic violence and abuse, life-threatening health conditions, and military combat. When a group of traumatized veterans was assessed to determine the impact of war on their clinical needs, a whopping 97 percent were in pain.[27]

The ACE Study was one of the first to show that trauma is unequivocally linked to health and disease. It has since been replicated hundreds of times across twenty-two countries, incorporating data from more than half a million people. These findings raise awareness about the impact of trauma and its long-term physical effects. They've also led to the development of interventions aimed at reducing the prevalence and impact of trauma on our health—strategies which you'll soon know, too.

But what hidden aspects of our neurobiology explain this puzzling pain-trauma overlap?

The Body Is the Bridge: The Neurobiology of Trauma

Trauma may be a social determinant of health, but it isn't just environmental, and it isn't just psychological. It also changes our bodies, disrupting the functioning of our nervous, endocrine, and immune systems. It also creates a neurological "imprint" that makes us more susceptible to developing chronic pain. Neuroscience studies reveal that trauma can alter brain pathways and neurotransmitters, rewire the amygdala, and send the pain system into overdrive.[28,29] Together, these neurobiological changes can create a heightened, intensified, and chronic stress response—which is why the "S" in "PTSD" stands for "stress." A nervous system stuck in survival mode continues to amplify danger messages even after the danger is gone, *turning the brain into a pain megaphone.*

On a more visible level, trauma also changes the way we think, feel, and behave. After a terrible, uncontrollable event, the world can suddenly feel like a dangerous place, leading us to become fearful and avoidant.

Living through one trauma inadvertently trains our nervous system to be on the lookout for others. If one terrible, unpredictable thing can happen, the brain reasons, why not another…? For example, someone violently mugged on the street may develop a fear of walking outside alone. A survivor of a school shooting may become wary of classrooms and crowded places. And someone viciously attacked by a dog may begin to dogmatically avoid them, no matter how friendly or small.

Lest we're tempted to believe this response is indicative of a defect or mental illness, it isn't. This is simply our brain doing its job exactly as designed: learning, associating, and adapting to protect us. Because we experienced a terrible, awful thing and weren't sufficiently protected, the brain goes into overdrive trying to do a better job next time. And the more traumatic experiences we have, the more overprotective our brains become.

A look at one hallmark trauma symptom, hypervigilance, shows these changes in action—and also illustrates their connection to pain. Hypervigilance is a hyperawareness of, and attunement to, small bits of sensory data. It occurs when an overprotective nervous system incessantly scans the environment for potential danger in a bid to protect us. As a result, our bodies hyperrespond to benign stimuli. A loud noise, for example, like a slamming door, can trigger an exaggerated startle response: jumping out of our chair, muscles tense, heart racing, ready to fight-or-flight. Hallie was all too familiar with this symptom, which kept her up at night.

But a hypervigilant nervous system doesn't just scan our external environment for danger; it also scans our *internal* environment, constantly on the lookout for potential threats. Because trauma makes the brain extra sensitive, small bits of harmless sensory data from the body can be interpreted as potentially threatening—triggering loud pain messages even in the absence of actual danger.[30] When benign bodily sensations are persistently misinterpreted as dangerous, this is a recipe for chronic pain.

Now. Not everyone who experiences a trauma will develop PTSD, just as not everyone who experiences trauma is doomed to develop chronic pain. But the overlap between pain and trauma has powerful—and surprisingly hopeful—implications for treatment. Because *treating one naturally lends itself to treating the other.* Indeed, research bears this out: Multiple treatments have been developed that can simultaneously reduce symptoms of both.[31,32]

Yet, despite all we've learned, trauma remains one of the most overlooked pain ingredients. To our great detriment, we continue to proceed with the dangerously reductive belief that trauma is simply a "mental health" diagnosis, and nothing more. In truth, we've known that the body keeps the score for a long, long time.

Tending to Your Garden

We don't always get to choose the soil in which we've been seeded: the cultures and contexts in which we've grown, or the hardships we've endured. However, while we don't have complete control over our early environments, there are many things we *can* control now. We may not be able to eliminate discrimination in medicine, but we can understand the health implications of bias and advocate for ourselves, as Serena Williams did. We can't do away with past traumas, but we can connect them to our symptoms and pursue effective treatments.

Ultimately, knowledge is power. Being aware of the elements in our environments that contribute to pain is essential—it finally gives us the agency to change them.

* * *

You have now been introduced to essential ingredients across the three pillars of pain: biological, psychological, and sociological. You've seen evidence that emotions, thoughts, attention, social factors, and environmental contexts each adjust the Pain Dial. This will ultimately enable us to take all of this information and harness it to reduce our *own* pain. The final, missing step? Seeing how these factors combine and work together to adjust the Pain Dial—an example of these interconnected cycles in action. How can the emotions we feel, the thoughts we think, the people around us, and the contexts in which we exist change our bodies and brain chemicals?

In order to demonstrate this, I will need the help of Hollywood.

9

The Body's Pharmacy

Movie star and acclaimed actor Michael J. Fox was a global phenom by the time he was twenty. But at the height of his explosive career, Michael suddenly fell off the map. Only years later did we discover why.

In 1991, at the young age of twenty-nine, Michael developed a debilitating, progressive illness called Parkinson's disease, wherein brain cells in the substantia nigra—a part of the brain that helps coordinate movement—inexplicably start dying. The death of these neurons leads to a severe deficit in brain levels of dopamine, resulting in impaired motor function, tremors, stiffness, and, at later stages, an inability to swallow. While Parkinson's has no known cure, it is currently treated with medications that increase brain levels of dopamine. Usually reserved for those in their twilight years, Michael's Parkinson's was labeled "early-onset." He graciously and generously let us into his life in a beautiful documentary called *Still*. His gait unsteady, hands trembling, he informs us that the scar on his face is from a fall he recently took in which his eye socket connected with an end table.

Exactly ten years after Michael's diagnosis, neuroscientists conducted a provocative experiment. They gave Parkinson's patients an injection of saline, an inert solution containing nothing but water and salt. The patients were not told this; they knew only that it *could* be saline, but that it could also be a powerful medication that would successfully reduce their symptoms. Remarkably, half the patients reported as much improvement after an injection of saline as Parkinson's patients typically experience while taking an actual medication.[1] And while astonishing, that was only the beginning.

When the patients' brains were scanned, scientists discovered that, in response to the mere expectation of improvement, their brains had started producing more dopamine—even though there was no drug. Not serotonin, not opioids—but dopamine, the very chemical they needed in order to heal.

The results weren't due to an error, and they weren't random chance. Instead, these same findings have been replicated across research with Parkinson's patients again, and again, and again.[2,3]

The Science of Self-Healing

Perhaps no better example exists of how the very ingredients we've been tracking—neurobiology, emotions, cognitions, social factors, and environment—work together to ease pain and symptoms than the placebo effect.

When we hear the word "placebo," we think it means an inert substance, a sugar pill—a nothing. In medicine, if a treatment is "no better than placebo," it is discarded as useless. But it's time to forget everything you've heard, because placebo is the opposite of nothing. The placebo effect occurs when something that *should* be nothing cures our pain, heals our symptoms, and makes us well again.

This phenomenon is also known by another name: self-healing. It may sound like hokey pseudoscience, but consider this: On a recent cycling trip, I hit some loose gravel and went over my handlebars. I was covered in lumps, bruises, and lacerations. Over the next few weeks, I watched as those black-and-blues faded to yellow, then disappeared entirely. And my skin, once scraped and bloodied, repaired itself so completely that it left no trace of injury. Just as our hands and knees can self-heal, so, too, can our brains. Fueled by neuroplasticity, the brain *constantly* repairs itself.

This is the reason we can awaken from a coma after a terrible accident, learn to walk again after a brain injury, and regain our ability to read and write after a stroke. In truth, we self-heal every single day of our lives. It's simply something we were designed to do.

One of the reasons we can do this is that our bodies come stocked with their own innate pharmacies, fully loaded with many of the very agents we need to heal. Among them: immune cells, clotting agents, feel-good brain chemicals, and self-repair mechanisms that target organs, systems, and biochemicals throughout the human body. There are many ways of activating this pharmacy to reduce pain via a combination of biological, psychological, and sociological ingredients. A placebo, believe it or not, is one of these ways. But in order to effectively harness its power, we first need to understand what a "placebo" really is.

Of all the things we've gotten wrong about placebo, the primary one is this: It isn't just one thing. It isn't just a sugar pill, and it isn't just our expectations. Rather, a placebo is made up of multiple different factors that combine to active our body's pharmacy, including many of the ingredients you've read about so far. Together, these create a "placebo effect," which occurs when our thoughts, expectations, and emotions—shaped by the social and environmental contexts in which they occur, and the neurobiological events they inspire—lead to beneficial health outcomes.

Research has now confirmed that placebos can activate our internal pharmacies to ameliorate the symptoms of a long list of very real diseases and ailments—including Parkinson's disease, Alzheimer's disease, migraine, heart disease, endometriosis, irritable bowel syndrome, cancer-related fatigue, anxiety and depression, and just about every kind of pain there is.[4,5,6] In fact, pain relief upon administration of a placebo is so legendary it even has its own name: placebo analgesia.

Amazingly, despite their elusiveness, placebos aren't rare or even hard to access. Instead, these self-repair mechanisms can be found in every doctor's appointment you've ever had, every medication you've ever swallowed, and every chapter you've read so far.

Sound mysterious? It did to Kai and Aiko, too.

A Secret Medicine for Kai

Kai, thirty-eight, an engineer at a prominent tech company, sat caved into his wheelchair, his back curved to the shape of the sagging fabric. He seemed defeated, barely looking up as his sister, Aiko, wheeled him through my office doorway. His hair was unwashed, his clothes wrinkled, his face gaunt and shadowed. His affect was flat, his face expressionless. He was heavily medicated, and his eyes were glazed and dull. But what stood out most were his feet. They were bare and filthy, with long, sharp claws that looked more like talons than toenails.

Kai had been diagnosed with Fabry disease as a child. Passed down from parent to child, Fabry is a rare, progressive disease that inhibits the production of enzymes that break down lipids, or fats. These lipids subsequently build up in blood vessels and tissues throughout the body, with disastrous results—including organ failure. On average, men with Fabry tend to survive only until their fifties. Their lives are often marked by pain.

Kai knew this all too well. After a debilitating episode of peripheral neuropathy—foot pain that felt like burning electricity—he struggled to stand, walk, or work. His toes had become so painful that he couldn't clip his toenails, let alone put on shoes. As he struggled to function, his lucrative career fell by the wayside. He became depressed and listless. That's when Kai and Aiko found me.

Within minutes of our first meeting, they shared with me that Kai had been told there was no hope for his condition. Some of the messages he'd been given, messages he'd believed and internalized, included "the nerves in your feet are permanently damaged," which birthed images of mangled, shredded nerve endings. He'd been warned that "cold temperatures would trigger foot pain," creating anxiety about cold showers and rain, and resulting in the avoidance of both. And finally, Kai had been advised that his symptoms would "only get worse with time," which convinced him that he was doomed to a life of ever-increasing pain.

Before I could do anything else, I needed to undo the damage that had been done. Because of the danger messages Kai had been given, his brain was constantly predicting worst-case scenarios. Even more damaging, Kai was convinced these outcomes were inevitable. We began the hard work of untangling beliefs and expectations from facts. We reviewed some Fabry research together—and discovered that cold water, touch, and low-impact movement, while potential triggers, were neither dangerous nor harmful to his body. Moreover, it was unlikely that "mangled nerves" were the sole cause of his pain, as no evidence of this had been found. In fact, Kai read, scientists believe Fabry pain crises result from a hypersensitive pain system launching into overdrive—not simply damaged toes. As such, I added a desensitization protocol to Kai's treatment plan to help his brain and body become less reactive. Each day, he walked for just five minutes. And every night, he placed his feet in a shallow pan of room-temperature water, while challenging beliefs with facts: "Water isn't dangerous, my body is safe."

Over time, as his predictions shifted and fear abated, Kai was gradually able to add cooler water to the pan, until he was able to soak both feet up to his ankles. He learned strategies to soothe his nervous system while doing the exposures, and they helped. As his brain desensitized and his body relaxed, his pain started to change. It was less intense and flared less often. Within weeks, to his astonishment, cool water no longer triggered pain. He still needed his wheelchair, but for the first time since his diagnosis, Kai felt optimistic. Maybe change was possible, after all.

When a doctor friend of Kai's suggested that CBD, or cannabidiol, a chemical found in marijuana widely marketed as a pain reliever, might help, Kai decided to try CBD gummies. They were cherry flavored and bear shaped. After the first dose, he became sleepy. Kai rolled his wheelchair into the living room and took a nap on the couch. When he woke, he was astonished to discover that his pain was gone. Could it be? He pushed himself up off the couch, incredulous, and tested his feet. First, he walked to the kitchen. No pain. Then, he walked to the end of the driveway. Still, no pain. He put on slippers, soft and forgiving, and took his feet on a short walk to the corner store to pick up some groceries. He waited for the familiar burning, the terrible prickling, the feeling of knives in his toes. But the pain didn't return. He was cured! Kai started calling friends to share the news, making plans to bike and ski. Sadly, it didn't last: By the time he woke the next morning, the pain had returned. He was again wheelchair-bound.

Still, it was an experiment worth repeating. Kai tried again the next day. Again, it worked: The gummies were followed by sleepiness, a nap, and then pain relief. This went on for weeks. Kai was delighted. His sister, Aiko, however, was suspicious: After everything she'd learned over the course of our work together, she didn't believe the CBD gummies were magically healing his disease and disappearing his pain. Could this effect possibly be from Kai's *belief* in the power of the gummies, she wondered, rather than the gummies themselves?

Aiko decided to try a secret experiment without telling Kai. She went to the store and purchased a bag of regular, nonmedicinal gummy bears. She placed them in the small jar Kai used to store his CBD gummies, and hid his original stash in the back of a drawer. When her brother wheeled himself into the kitchen that afternoon, she handed him a regular cherry gummy bear—without telling him about the switch. Aiko waited for Kai to complain that the CBD wasn't working. Instead, to her surprise, the usual happened: Kai became sleepy, rolled himself into the living room, and fell asleep on the couch. An hour later, he woke, broke into a delighted smile, and pushed his wheelchair aside. His pain was gone. He put on his shoes, waved goodbye, and left to walk to the corner store.

Aiko stared at the door as it closed behind him, her eyes wide and full of wonder. She fed Kai regular cherry gummies for an entire week, always certain that the next day would be the day the experiment failed. But it never did. Instead, the placebo gummies continued to work. Every day, Kai's pain disappeared—he got out of his wheelchair, went for a walk, and resumed his life until the placebo effect wore off.

At the end of seven days, Aiko, ashamed, broke down and told her brother what she'd done. She awaited his justifiable wrath. Instead, Kai embraced her. The truth was confirmed: While his disease was a fact of life, his pain didn't necessarily need to be. In fact, it was treatable. And medications weren't the only solution. Instead, there was a remedy he'd been carrying inside him all along.

Accessing the Pharmacy: The Science of Safety

Only pain science can make these baffling findings make sense. Over the past few decades, researchers have conducted a deep dive into the human brain to uncover what mysterious processes could possibly underlie stories like Kai's. Was it purely psychological? Or was there a chemical, neurobiological explanation? The answer, as you've likely guessed, is both—and more.

The ingredients in a successful placebo vary, but overall, they're remarkably consistent. Across studies, the medicinal factors that activate the body's pharmacy to reduce pain include:

1. A treatment—real or fake—like Kai's placebo gummies.
2. The rituals around that treatment's administration—for example, a common medical ritual like swallowing a pill with water.
3. A credible, trustworthy source—for example, the scientific papers Kai read, me, and Kai's doctor friend who recommended the gummies.
4. The specific words and messages offered by that credible source—for example, the doctor's assertion that the gummies could offer pain relief.
5. Positive expectations, beliefs, and predictions.
6. Positive emotions, like hope, trust, and optimism.
7. A safe, soothing environmental context.

If we take a closer look at the elements that activate our pharmacy to create a placebo effect, a clearer picture emerges. Let's begin with the world around us. The environment in which a treatment is delivered, for example, has a substantial impact on how well a treatment works—beyond what we ever imagined.[7] We're more likely to experience pain reduction in a credible environment that inspires feelings of safety, like a doctor's office or hospital, than in a space that feels unprofessional, unsafe, or suspicious. (This is why so many of us have had the mind-bending experience wherein pain inexplicably vanishes the instant we arrive in our doctor's waiting room: Environmental safety messages immediately soften the pain alarm.)

Social factors matter, too, like verbal and nonverbal cues from the people around us. Someone wearing a white coat and stethoscope, a uniform we've come to associate with medical authority and wellness, is more likely to win our buy-in and change our emotions and expectations than someone wearing a FedEx uniform or a clown suit.

The emotions inspired by these people, and the words they choose to use, also affect our pharmacies. If the person administering a treatment is warm and empathic, expresses confidence in the treatment, and inspires positive emotions like hope, trustworthiness, and optimism, we're more likely to experience pain relief than if messaging makes us feel hopeless, scared, or discouraged. Finally, our own predictions and expectations—like Kai's belief that the gummies would reduce his pain, and my patients' belief in our work together—also influence our body's pharmacy. If we *believe* a treatment is going to work, it's more likely that it *actually will*.

When these factors come together, something incredible happens inside our bodies. The brain initiates a cascade of physiological events, coordinated in large part by our prefrontal cortex—the frontmost part of our brain that plays a key role in making meaning and predictions.[8,9] The prefrontal cortex synthesizes all available data from our body and our environment, translates this into chemical and electrical signals, and shares it with other brain sites and body systems. This, in turn, triggers physical changes in our blood flow, heart rate, neurotransmitter levels, and hormones. It alters immune response, the functioning of our nervous system—and it changes pain perception.

If we zoom out and examine this information from an eagle's-eye view, a familiar theme emerges, one you've seen before: the theme of danger and safety. Over and over again, science shows us that these play a major role in pain perception. So, how does this work? When factors combine to induce feelings and expectations of *safety*, the body's pharmacy is activated. These "safety messages" alter activity in the parts of our brain and spinal cord that make pain, lowering the pain alarm.[10] They also affect our brain chemistry, inducing changes in our opioid, serotonin, dopamine, and endocannabinoid systems. If we'd been able to conduct a scan of Kai's

ACTIVATING THE BODY'S PHARMACY

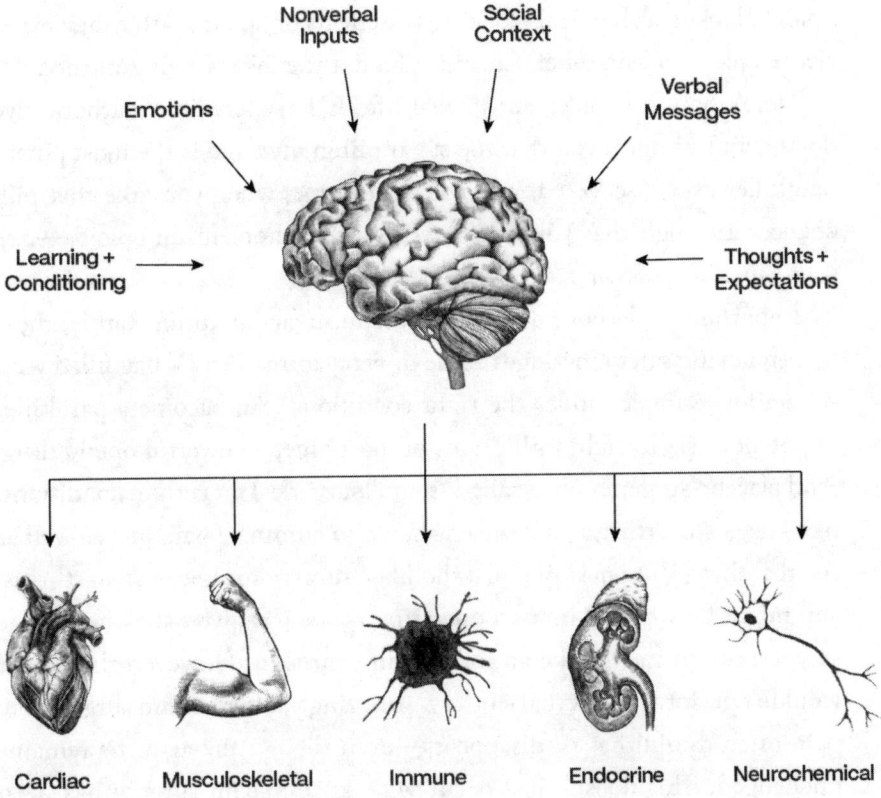

The body's pharmacy is activated by a combination of biopsychosocial inputs that affect cells and systems throughout the human body.

Credit: Illustration by Eva Huzella

brain after he took that placebo gummy, we likely would've observed these very changes. Homemade opioids play a particularly important role here. One of the ways scientists have tested this is by giving research subjects an opioid blocker, a chemical blocking the effects of opioids, after they experienced placebo pain relief. Lo and behold, the subjects' pain returned.[11]

Here's how this plays out in real life: If I'm a credible, authoritative doctor, and I inform you that the sugar pill in my hand is the most potent painkiller ever discovered, and you believe me, when you take that pill, chances are high that your brain will produce homemade opioids—*and your pain will go down.*

Importantly, placebos don't just come in sugar-pill form. Rather, these pain-relieving safety messages come in *many* forms. An IV bag filled with saline, for example, under the right conditions, can become a painkiller as potent as six to eight milligrams of morphine, a powerful opioid drug. And placebo surgeries (also called "sham" surgeries) for certain conditions, like knee osteoarthritis, have been shown to eliminate pain just as well as the real thing.[12] As part of a placebo knee surgery for knee osteoarthritis, you might be wheeled into an operating room and given anesthesia. The surgeon might then make an incision in your skin. However, the doctor wouldn't perform any actual surgery. Amazingly, after a sham surgery, the pain often diminishes or disappears—even though the arthritis remains unchanged. This doesn't just occur with knees: sham surgery has been repeatedly successful at reducing pain across a wide swath of conditions and body parts.[13,14]

Ultimately, scientists have concluded that the rituals and context around these surgical procedures for chronic pain, including the predictions and emotions they inspire, are *just as important for pain relief as the procedure itself.* This is why the mere act of receiving treatment—any treatment, whether an injection of saline or a bogus surgery—can, under the right circumstances, be effective in and of itself.

Now. This isn't to suggest that surgery isn't important. Surgery is critical for repairing torn tissues, removing dangerous growths, extracting rotting teeth, and treating malfunctioning organs, among other things.

In the case of dangerous conditions like cancer, surgery can be lifesaving. This also doesn't mean that sugar pills should replace medical treatments for damage or disease. That's as dangerous as it is false. Placebos cannot mend broken bones, cure cancer, or eliminate Fabry disease. (While Kai's pain was less, his inherited disease remained.) Moreover, placebos don't outperform all medications for every illness and condition—in nearly all cases, superior drugs are available, including for Parkinson's disease. Nor should providers deceive their patients, or swap out pain medications for saline during surgery. Administering a faux treatment correctly and well requires training, preparation, carefully constructed cues, and—most important—informed consent.

However, when it comes to pain—the *hurt*, not the harm—the implications of this research are enormous. Effective pain treatment doesn't necessarily require cutting, ablating, or drugging our bodies. Because our body's pharmacy has the power to generate some of the very remedies we need to heal. And we can activate it with the right mix of ingredients.

If this seems too good to be true, take heed: A different combination of factors can do the exact opposite. As we saw with Kai at the beginning of his treatment, negative expectations and beliefs have the power to hijack our body's pharmacy and make us feel worse. Because just as words can heal, they can also harm.

Black Magic, Voodoo Medicine, and Nocebo Science

Voodoo, curses, hexes, evil spirits, and black magic pervade every major religion and ancient culture. But they don't just live in spell books. Respected medical journals have been reporting on mysterious deaths at the hands of curses and evil spirits for years—deaths considered particularly remarkable because the victims had no injury, illness, or infection. These incidents have come to be known as "voodoo deaths," psychogenic deaths, or psychosomatic deaths.

Consider the alarming case of Mr. A, age twenty-six.[15] Arriving at the emergency room, he approached the front desk and cried out, "Help, I've taken all my pills!" As he collapsed to the floor, he dropped an empty prescription bottle. Mr. A was rushed to an exam room for assessment. He was drowsy, lethargic, and pale. He reported that he'd been prescribed an experimental drug for depression by doctors at a local university. The label on the bottle confirmed that it contained capsules that were part of a clinical trial for antidepressants. Feeling suicidal, Mr. A had consumed the entire bottle, all twenty-nine capsules, intending to end his life. He regretted his decision, he said, and hoped to be saved.

But his symptoms were dire. He was hypotensive, his blood pressure bottoming out. He was faint, sweating, and trembling. His heart rate was elevated, and he was having trouble breathing. Mr. A required intravenous fluids to maintain adequate blood pressure. He seemed to be slipping away. The ER team conducted a battery of tests to determine the nature of his overdose, and what treatments were needed to save him.

When the results came back, the team was bewildered...and then stunned. Every single test was negative. All laboratory studies were normal. In fact, there was no trace of any drug in his body at all. After consulting with the researchers running the clinical trial, Mr. A's doctors concluded that he'd never taken a medication. Instead, Mr. A had "overdosed" on sugar pills. When his ER team revealed this to him, Mr. A wept with relief. He wasn't going to die, after all.

Within fifteen minutes, his symptoms were gone.

In recent years, science has been able to explain the neurobiology of these medical mysteries. When our thoughts, expectations, and emotions—shaped by the social contexts and environments in which they occur, and the neurobiological events they inspire—negatively impact our health, this is known as a "nocebo effect." "Nocebo," which comes from Latin's *nocere*, meaning "to harm," is the opposite of "placebo": a *danger message* with the power to trigger deleterious changes across the body and amplify the brain's pain alarm.[16,17]

The Anatomy of Danger Messages

What was the difference between Kai's faux "CBD gummies" and Mr. A's "antidepressants"? Answer: Everything the brain used to infuse them with meaning. The social and environmental contexts surrounding them, the emotions and physical experiences they produced, the expectations and predictions they inspired, and the resulting neurobiological events.[18] *Bio-psycho-social.*

In the case of Mr. A, the nocebo effect consisted of cognitive factors—his belief that those pills were a real drug, and the expectation that they'd cause his death in overdose. It included negative emotions like panic, fear, and stress, which contributed to physical symptoms like sweating, elevated heart rate, and blood pressure changes. Social and environmental contributors included the doctors that prescribed the pills; the ritual around swallowing them with water, which our brains have come to associate with taking real medications; and the university setting that gave the treatment credibility. It included messages, like the doctor's announcement that Mr. A and the other patients were potentially being prescribed powerful antidepressants.

Simply put, a danger message is significantly more dangerous if we believe it; if it generates negative predictions and expectations; if it inspires fear, anxiety, or dread; and if it's delivered by a credible source, like a healthcare provider.[19] This, scientists believe, is how voodoo curses and even sugar pills can sometimes, under certain circumstances, actually kill their victims. And if nocebos can cause death, rest assured they can trigger and amplify pain. Research confirms this is indeed true.

Clinical studies on the nocebo effect are relatively rare—thankfully, it's unethical to deliberately hurt people in medicine. But scientists have managed to explore this phenomenon in some creative ways. In one revealing study, subjects voluntarily held a hot device while researchers cultivated expectations using words and images. Participants were warned of intense heat with a red sign that read "high temperature." The researchers picked red deliberately: In nature, red is a sign of danger, the ominous color of some of the world's most poisonous animals, like red widow spiders and coral snakes, and is thus a color our brains are primed to associate with harm. Expectations of low heat—and, subsequently, low or no pain—were induced using a blue sign reading "low temperature."[20]

As hypothesized, expectations, emotions, messages, and social context significantly altered the amount of pain people ultimately perceived. Participants consistently rated a given temperature as more painful when they expected high heat—confirmed by visible, measurable activity in the

Maximal activation of the brain's pain circuitry was found only when high heat was paired with *expectation* of pain.

Credit: Image from Keltner J., Furst A., Fan C., et al. (2006). Isolating the modulatory effect of expectation on pain transmission: A functional magnetic resonance imaging study. Journal of Neuroscience, 26(16), 4437–4443. Copyright 2006, Society for Neuroscience.

brain's pain circuits. More shockingly, it wasn't high heat alone, even at its most intense, that triggered the biggest pain response. Rather, maximum activation of the brain's pain circuitry was found only when high heat was paired with *expectation of pain*.

Hidden in Plain Sight: Nocebos in Everyday Life

Nocebos occur in everyday life, and more frequently than we might think. That unforgettable construction worker with the nail in his boot (but not his foot)? His pain was triggered by a "nocebo recipe" composed of all the critical ingredients. As he jumped off the plank onto that nail, his brain gathered all available information—including knowledge of his dangerous work environment, memories of past injuries, his coworkers' horrified faces, the sight of a seven-inch nail piercing his boot, and the *belief* his skin had been punctured. Because there was sufficient evidence of potential danger, despite the lack of actual damage, his brain made pain to protect him.

I see the effects of nocebos in my clinic every day, mostly in the form of insidious messages unintentionally delivered by well-meaning providers, family members, and media. I'm hard-pressed to think of a single patient, including myself, who hasn't been subjected to them in one form or another. Hallie was informed by the internet that fibromyalgia was "incurable" and that her pain was permanent. Joyce the dancer was told she had "the back of an eighty-year-old" and that her pain was due solely to damaged, degenerating discs. Kai's belief that his "mangled nerves" were the sole cause of his pain, rendering him untreatable, was a nocebo, too.

Nocebic messages are everywhere, hidden in plain sight.

There's evidence that the damage done by common examples like these isn't merely anecdotal. A collection of studies found that inert substances can trigger migraine attacks when patients are told to expect one. Injections are more painful when we're warned we'll feel "bee stings" and "burning" compared to a "numbing sensation."[21] People with low back pain who were given negative messages by their physiotherapists, such as "your muscles are weak," were more anxious and held more pessimistic beliefs than those who received none.[22] Recent research reports that 90 percent of the muscle pain purportedly caused by statins—a medication commonly prescribed for high cholesterol—is actually due to a nocebo effect triggered when we're told to expect muscle pain, and not the drug itself. Indeed, we've long known that telling someone they're likely to develop a side effect after taking a medication notoriously makes it more likely to actually occur. Amazingly, this is true even if the side effect is completely made up.[23]

"I'm not so much concerned with the side effects of the drug I gave you as I am with the fact that it was a placebo."

Credit: Illustration by Jon Carter/Cartonstock.com

Media reports publicizing and sensationalizing negative reactions to medications have even been known to trigger public outbreaks of adverse side effects. In 2018, for example, news stations in New Zealand reported that a popular antidepressant triggered side effects like nausea, brain fog, headaches, and suicidal thoughts. Immediately after the stories aired on TV, the number of adverse side effects reported to the New Zealand Centre for Adverse Reactions Monitoring increased by a whopping 4,283 percent.

Panic-inducing news reports can even have a nocebic effect on an entire global population, as many hypothesize occurred during the COVID-19 pandemic, triggering and exacerbating adverse reactions to vaccines. Negative prognoses and diagnoses can have a negative health impact, too, quickening the arrival of poor outcomes and worsening symptoms. A study published in the *New England Journal of Medicine*, for example, reported that simply telling a patient they have cancer can, under certain circumstances, be enough to trigger a heart attack.

This isn't to suggest that our care providers should stop warning us of side effects, informing us about our diagnoses, explaining prognoses, or expressing concern. These are critical parts of good, ethical medicine. However, the language we use *matters*. Well-intentioned providers and family members may share unfortunate or discouraging statistics, like low survival rates, high rates of relapse, or dreadful side effects. They may use catastrophic language or make negative predictions to help us manage expectations. And we might spend time looking up our symptoms on the internet, reading horror stories, and thinking terrifying thoughts. But while these urges are perfectly normal, they don't actually help us. Instead, they hurt us. Because danger messages trigger harmful expectations, predictions, and emotions that hijack our bodies, amplify pain volume, and make us feel worse.

Hope on the Horizon: Taking Control

There are some fresh takeaways here that add to our growing understanding of pain. One is that we have the ability to self-heal—in no small part because our bodies come with their own well-stocked pharmacies. The second is that there are many ways of activating this pharmacy to reduce pain via a combination of biological, psychological, and sociological ingredients—placebo being the perfect example. And the third is that nocebos, which can come in the form of messages from the outside world and even the thoughts we think, are part of every high-pain recipe. Nocebos show us how these ingredients can work together to *increase* pain and symptoms. Because of how common they are, and how hidden, we need to learn how to identify and manage them—something I will teach you how to do.

We've also added an additional ingredient to our pain recipe: words. The words we hear, the words we read, and the words we speak inform our thoughts and emotions. Our thoughts and emotions inform our physical sensations. In this way, the messages we believe become the things we predict, feel, and perceive. If these ingredients sound familiar, you met them once before in Chapter 6: They are the fundamental pieces of the Pain Cycle.

When it comes to pain, understanding how all ingredients work together gives us the chance to exert agency. Because we can learn to identify nocebos and danger messages. We can notice the suggestions, both positive and negative, fed to us by friends, the internet, social media, and healthcare. We can decide which to reject, and which to believe. We can carefully choose the words we use, and challenge the thoughts we think. We can pay attention to our beliefs, and monitor our expectations and predictions. As providers, we can avoid fearful, catastrophic predictions, convey faith in our treatments, and use thoughtful language.

There is another silver lining here, too. This new knowledge gives our healthcare providers the opportunity to reduce use of nocebos while safely, ethically administering "safety messages" in the form of words, actions, nonverbal cues, and even faux procedures that have the power to heal. A great template for effective use of open-label placebos—that is, fake treatments that the patient knows are fake, but work anyway!—already exists at Harvard Medical School's Program in Placebo Studies. The research emerging is astonishing: Open-label placebos are effective not just for "psychosomatic" illnesses, but also for very real ones. Consider the curious case of a disease called irritable bowel syndrome, or IBS.

IBS is an unfortunately-named, common condition characterized by severe abdominal pain, bloating, constipation, and all manner of gastrointestinal upset. It can last for decades and has no known cure. It wasn't entirely surprising, then, that eighty adults with IBS found themselves registering for a Harvard study researching an unusual treatment: a placebo pill. Unlike other placebo studies, participants were told up front that the pill was a placebo. They were simultaneously informed that this very same placebo had successfully treated many others with IBS, considerably reducing their pain and symptoms. The patients, understanding the terms, took the pill twice a day, gulping it down with a glass of water. For the duration of the study, they were given attention, support, and information from trained medical professionals.[24]

Then, they waited.

After three weeks, 60 percent of the patients felt better—commensurate with the improvement rates achieved by popular IBS drugs.

And it's not just IBS. While the field is young, open-label placebos have similarly demonstrated effectiveness for treating chronic low back pain, postsurgical pain, stomachaches, cancer symptoms like Coach Murph's, and can even reduce our need for opioids.[25] Imagine if we all learned how to activate our body's pharmacy—to self-administer this real, free pain medication available everywhere, from our homes to the ER? What a powerful, masterful army we would be!

To this end, some research institutions are already making incredible strides toward helping us maximize this untapped potential. These studies investigate a range of things: the biochemical bases of placebo, the brain circuits it activates, even its effectiveness for specific conditions like Parkinson's disease. Recently, a $12 million grant was awarded to neuroscientists studying the impact of placebo on Parkinson's, with the aim of speeding the path to a cure. Who donated this phenomenal grant, you ask?

The Michael J. Fox Foundation, of course.

Many Things Are Medicine: A New Approach to Healing

Two billion of us around the world live with pain. Yet the myth persists that only a handful of treatment options exist.

The truth is that medicine comes in many forms. "Pain medication" isn't just pills, although those can certainly help. It isn't just surgeries and procedures. True pain medicine is any strategy or intervention that transforms a high-pain ingredient into a low-pain one. The things we think, the things we feel, how much we sleep, what we eat, the messages we receive, and even the people we interact with adjust our Pain Dials, too. This means that there isn't just one way to treat pain—

There are *a million*.

You've now seen proof that together, biological, emotional, cognitive, social, and environmental factors constantly change the pain we feel. Even if we can't eliminate our pain entirely, we can tweak these ingredients to change its volume. This is incredible information. It means that, rather than being at pain's mercy, we have the power to break the Pain Cycle.

In Part III, I'll give you actionable, easy-to-follow guidelines for doing exactly that. I'll outline a science-backed pain protocol with proven effectiveness, one that focuses on *all of you*: your brain, heart, blood, and bones. I will teach you how to harness the power of the ingredients you've discovered here to create your own low-pain recipe, stimulate neuroplasticity, and activate your body's pharmacy.

Together, these tools will help you:

- reduce pain intensity and frequency
- break the Pain Cycle and activate the Healing Cycle
- rewire and desensitize your pain system
- increase strength, physical performance, and mobility
- increase hope and resilience
- make better nutritional choices to reduce pain and inflammation
- improve sleep
- boost mood and reduce stress

- increase social support
- shift internal dialogue and self-talk to best support recovery
- improve quality of life, and
- speed healing.

You've seen this approach work for Sam's migraines, Mateo's phantom limb pain, Joyce's back pain, Fallon's complex regional pain syndrome, Sarah's abdominal pain, Kiran's kidney stones, Coach Murph's cancer pain, Hallie's fibromyalgia, Kai's Fabry disease.

And now, it's time for me to teach you.

PART III

The Pain Protocol

10

Welcome to the Revolution

FROM PAIN TO POWER

You thought this was a book about pain. It's also a book about hope.

To say that the current state of pain medicine is "inadequate" is an understatement; that much is clear. And in the wake of the opioid epidemic, we've been left desperate for answers. Fortunately, science has them.

And now you have them, too.

Parts I and II of this book have revealed the truth about pain, finally rescuing it from the dusty textbooks and medical journals in which it has been buried for so long. That isn't where science belongs. It belongs to all of us—to you, to my patients, to me. What this science tells us is that chronic pain is treatable, that we can mend and heal. However, until our healthcare system is revamped and this science is widely disseminated, change has to start with us. This is the pain revolution, and you are now part of the clarion call.[i]

In Part III, you'll take what you've learned and put it to work, harnessing the power of neuroplasticity and your body's ability to self-heal. Together, we'll craft your Pain Protocol, a road map for treatment that

---

[i] Credit to the great Dr. Lorimer Moseley, friend and colleague, for pioneering and popularizing the term *pain revolution*. His Pain Revolution, based in Australia, is a powerhouse pain education-and-treatment program worthy of emulating. Find his work at painrevolution.org and tamethebeast.org.

will finally allow you to take control of the dial. This section offers a guide to the many available treatments for chronic pain, solutions you can start implementing right now. I'll introduce you to practical, effective, science-backed strategies for improving functioning, enhancing health and quality of life, and reducing your pain. I'll break down each recommendation into an action plan so that you have a step-by-step approach, and offer tips for finding providers to support your healing journey.

My patients who follow this protocol get out of bed and back to life. And they're not alone. Research shows that this approach, one that incorporates all pieces of the pain puzzle, is the most effective—well beyond pills or procedures alone.[1,2,3,4,5,6,7,8,9,10] The secret to treating chronic pain isn't a well-marketed, "revolutionary" pink pill. It isn't a newfangled technique, and it isn't out of anyone's reach. The true treatment for chronic pain is biopsychosocial medicine—the very medicine I offer you here.

And it works if you're willing to work it.

Of course, this approach, like any, isn't without its challenges. For one, pain treatments beyond pills and procedures are frequently denied insurance reimbursement. I address this by offering suggestions to make all parts of this protocol both accessible and affordable. Second, pain medicine has become so siloed, its component parts so fragmented, that you've likely seen at least a handful of providers to tackle your pain problem—most of whom have never spoken to each other, and never will.[ii] We saw this with Fallon, who had a team of nine healthcare providers, each of whom targeted a different body part. We also saw this with Sam, whose physical health had been so divorced from his emotional health that his emotions were simply ignored. Part III will address this by outlining interventions that span these disconnected disciplines—from medicine to physical therapy to psychotherapy—finally integrating the three pillars of pain we need to target if we want to get well.

While much of this book is directed toward people living with pain,

---

ii This isn't the fault of clinicians, but rather a byproduct of various external factors—including the impossible demands on clinicians' time, lack of reimbursement for care coordination, infrastructure challenges, lack of standardized communication channels, incomplete disclosures by patients, among others.

I also dedicate sections to those who treat pain. These sections present opportunities for healthcare providers of all backgrounds to integrate their training and knowledge with the best of other disciplines—a place for doctors, physical and occupational therapists, mental health professionals, nurses, and others to come for invaluable resources, tips, and tools.

A How-To Guide for Healing

The following section is divided into five parts. These parts map onto the topics outlined in Parts I and II, yielding the five critical components of your Pain Protocol:

1. How to identify and track your pain recipe.
2. Biological interventions targeting sleep, nutrition, movement, activity, and functioning.
3. Strategies targeting emotional health, including brain-based treatments known to adjust pain volume.
4. Cognitive techniques to modify negative thoughts and nocebic messages and master the skill of distraction.
5. Social medicine, with a special focus on healing trauma, setting better boundaries, and creating healthy social connections.

Diagnosing and treating disease and damage with your team is also a critical part of this protocol, a step you've undoubtedly already taken. As you proceed, keep in mind that changing just one part of your Pain Cycle affects *all* parts. As the cycles throughout this book have shown us, the pain process is so interconnected that improving how you eat, sleep, and move will inevitably change how you feel physically and emotionally; changing how you feel physically and emotionally will impact your relationships and social health; and changing your social health will alter your biology and psychology. Ultimately, by addressing all the ingredients in your pain recipe, this protocol targets all of you—head to toe and inside out—from the pain pathways in your brain and nervous system, to the blood and immune cells that speed tissue healing.[11,12]

As a first step, grab a notebook or open a computer document in which to take notes. Here you will track your treatment protocol, log your pain recipe, and answer the questions posed in this next section. Once you have this prepared, you're ready to begin.

A WHOLE-PERSON APPROACH TO PAIN

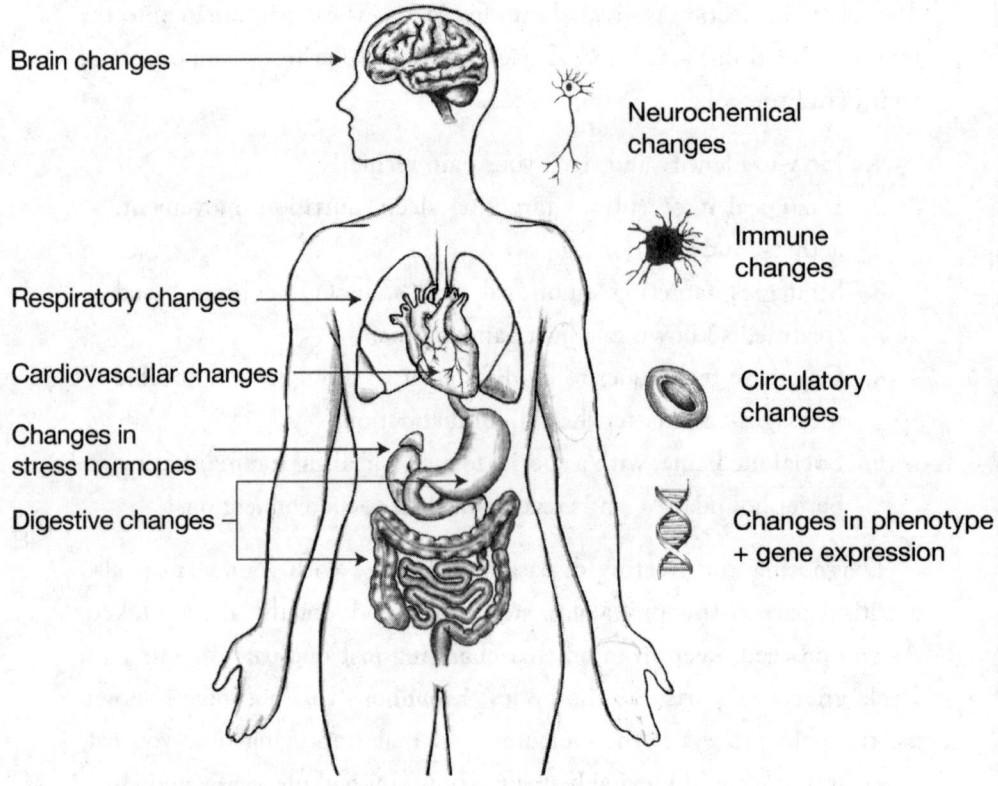

This pain protocol targets all of you—from the pathways in your brain to the immune cells that speed tissue healing.

Credit: Illustration by Eva Huzella

Crafting Your High Pain Recipe

By now, you may have identified some of the ingredients in your high-pain recipe: the various factors that collude to trigger and amplify your pain. You've likely seen some of these ingredients reflected in the characters you've met in this book. But some may still be hidden. That's to be expected. To help uncover these, begin by answering the following questions in your notebook:

- *When your pain feels worse, what triggers the flare?*
- *When your pain feels less bad, what is it that helps?*
- *Have you noticed any patterns to your pain?*
- *Do you tend to feel worse when you sleep poorly, skip meals, or have a particularly intense workload?*
- *Is pain less intense on certain days of the week, or certain times of the year? For example, some of my patients tend to have flares on Sunday nights and Monday mornings—an indication that work or school stress is a likely high-pain ingredient. Others notice that pain worsens during winter months or when it rains, and dials down with the summer sun.*
- *Have you ever gone on a trip or vacation and noticed that pain volume dialed down? If so, what stressors were absent during those periods that are normally present? What factors were present that are normally absent?*
- *Think back to the time your pain started. Beyond injury or illness, what else was going on in your life at that time?*

To support your investigation, I've included a list of some common high-pain ingredients in the menu below. Select the ingredients that apply to you, and log them in your notebook. In the diagram on page 178, you can group these according to their category (bio, psych, social) using the blank template provided. For example, if poor sleep is a pain trigger and amplifier, list it under "bio." If you've been struggling with stress or anxiety, log that under "psych." If you're navigating divorce, family conflict, or the death of a loved one, add that under "sociological," which combines social and environmental factors. As you make your way through Part III, continue to add any high-pain ingredients you identify to your list. Soon, a clearer picture of your own pain recipe will begin to emerge.

You may find that your recipe has more ingredients than you realized, or that some are missing from the list I've provided. Alternatively, you may have trouble immediately identifying all of your high-pain ingredients. Have no fear—we'll thoroughly review each pillar of pain in the coming chapters, digging into the specific factors that fall into each. After completing Part III, you'll be able to identify many more contributors—and still more in the coming weeks and months. This is a *good* thing: Every ingredient you uncover is an opportunity to change your recipe.

HIGH PAIN INGREDIENTS

Biological ingredients
_____
_____
_____

Psychological ingredients
_____
_____
_____

Sociological ingredients
_____
_____
_____

HIGH PAIN MENU

Consider ingredients in this menu that apply to you, then add them to the list in the blank template above.

Bio	Psych	Sociological
Tissue damage, system dysfunction, disease, illness	Unhealthy coping, like drinking, isolating, or avoiding	Trauma
Inflammation	Kinesiophobia (fear of movement)	Toxic or unhealthy relationships, divorce, death of a loved one
Genetics, hormones, neurochemicals	Nocebic beliefs and predictions: expecting dangerous, negative, or bad outcomes	Environmental stressors: work, family, financial stress; social media; politics; deadlines; bills; etc.
Highly sensitive nervous system	High stress	Isolation, insufficient social support
Insufficient or poor sleep	Interpretating sensations as dangerous	Unsafe or unstable living environment
Poor diet, inadequate nutrition	Emotion suppression, "shoving it down"	Nocebic messages from the outside world (media, family, friends, healthcare providers)
Being sedentary, avoiding movement or activity	Anxiety, depression, other mental health issues	Poor boundaries
Muscle tension	Negative thoughts	Inadequate access to care
Overreliance on pain medications, alcohol, or other substances	Spending a lot of time focusing on, thinking about, or talking about pain	Racism, inequitable care

Your Low Pain Recipe

Constructing a high-pain recipe is both instructive and illuminating—because it points us to our low-pain ingredients. These are easily identified using this method, because they are usually the *opposites* of your high-pain ingredients. For example, if insufficient sleep is a pain trigger and amplifier (high-pain ingredient), our low-pain ingredient is better sleep. If stress, anxiety, and muscle tension perpetuate pain, the low-pain ingredients are relaxation and stress reduction.

Your next step is to grab your notebook or laptop and create a table with two columns that looks like my Sample Pain Recipe, below. As you craft yours, you can also refer to Joyce's back pain recipe from Chapter 4. First, enter your list of high-pain ingredients on the left. Then fill in your low-pain ingredients on the right. Seeing them laid out before you in black and white will prove to be powerful medicine.

SAMPLE PAIN RECIPE

High Pain Recipe	Low Pain Recipe
Poor sleep	Quality sleep
Trauma history	Trauma treatment
Being sedentary	Activity and movement
High stress	Stress management
Nocebic danger messages	Safety messages
Inadequate nutrition	Dietary changes
Inflammation	Medication, ice, dietary changes, stress management, exercise

The Transformation: Strategies That Work

This is all critical data. But there's a missing step, a yawning gap between the problem and the solution: What strategies must we employ to transform a high-pain ingredient into a low-pain ingredient? What are the methods and tools needed to improve sleep, treat trauma, address nocebic messages, and lower the pain alarm? In the coming chapters, I'll offer you interventions to transform your pain recipe, along with the science and evidence behind why they work.

More information about these strategies, along with a collection of additional tips and tools, can be found in *The Pain Management Workbook* (published by New Harbinger Press) and its Spanish translation, *Gestiona Tú Dolor* (published by Editorial Sirio). These workbooks contain an affordable, accessible treatment protocol that can be used by anyone in pain regardless of diagnosis, and by any healthcare provider of any training or background. Additional book recommendations, journal articles, podcasts, educational videos, guided audio, apps, and other zero-cost resources can be found at Zoffness.com.

Two Steps Forward, One Step Back

If your path has been a nonlinear, frustrating one, hang in, because you're on the right track. Healing from pain isn't typically a straight line toward recovery, with every day arriving better than the one before. The expected road to recovery from chronic pain is "two steps forward, one step back." Flares and setbacks are a normal part of the healing journey—not a sign of failure, and not a sign that your body is betraying you. This is just how our brains and bodies work.

This means that, as we construct our pain recipes and establish protocols for healing, we should *expect* that progress may be punctuated by flares and periods of little progress. Try not to get discouraged or give up hope. Planning ahead and becoming familiar with the forward-backward-forward process reduces uncertainty, frustration, and fear, and helps us move forward with confidence. Part III is where your new knowledge becomes actionable—and your pain recipes become interventions. This agency is at the very heart of healing, and your new knowledge will help chart your course.

II

The Biology of Balance

SLEEP, DIET, AND MOVEMENT

A biopsychosocial approach to pain never means ignoring the biological domain of pain. *Not ever.* If your healthcare providers don't work together to ensure nothing dangerous is happening to your bones, blood, and body parts, they haven't done their job. But addressing tissue damage and disease is only the first step. In this section, we'll explore proven strategies for adjusting the biological domain of pain by targeting:

1. when and how we move, and how much
2. engagement in activities and hobbies
3. the amount and quality of our sleep
4. the fuel we put into our bodies, and
5. pain medications.

Together, these strategies help us achieve and maintain physiological balance, a process known as "homeostasis." This is where our path to healing begins.

Homeostasis: Regaining Balance

The biological need for homeostasis is so fundamental to our health that our bodies constantly send us messages when we're off-balance asking us to restore it. Hunger pangs and headaches, for example, are signals we're low on fuel and need to eat. That chill that makes you shudder? That's your brain urging you to bump the heater or grab a blanket. If you sit in one position for too long, endangering tissues and tendons, your body will instruct you to stand, stretch, and move. Our bodies have many ways of telling us that something's off-balance—including physical symptoms like pain. And that pain urges us to attend to the neglected ingredients in our recipes that might restore balance—*and turn down pain volume*. One of the most important among these is movement, a critical ingredient on our path to recovery.

I. Retraining the Pain System with Activity

A. A Body in Motion Stays in Motion, a Body at Rest Stays at Rest

As we've seen, one reasonable response to pain is to avoid anything that triggers it, including movement. We can even develop a fear of moving our bodies when we have pain. The term for this is "kinesiophobia," and it stems from a concern that activity will make a painful condition worse.[1] While these "stop" and "avoid" instincts are adaptive and even lifesaving when it comes to acute (short-term) pain, preventing us from running on that broken leg and ensuring we rest when sick, this approach to healing doesn't work for chronic pain.[2]

Ultimately, this avoidance is a trap. Avoiding movement and exercise creates a toxic cycle of immobility, disability, and pain over time as our muscles atrophy, joints stiffen, bodies weaken, mood crashes, and pain system sensitizes. We can't move because we're still in pain, and we're still in pain because we're not moving. The brain, stuck in danger mode, never has a chance to learn that the loud pain alarm is no longer needed—that *some movement is actually safe*.

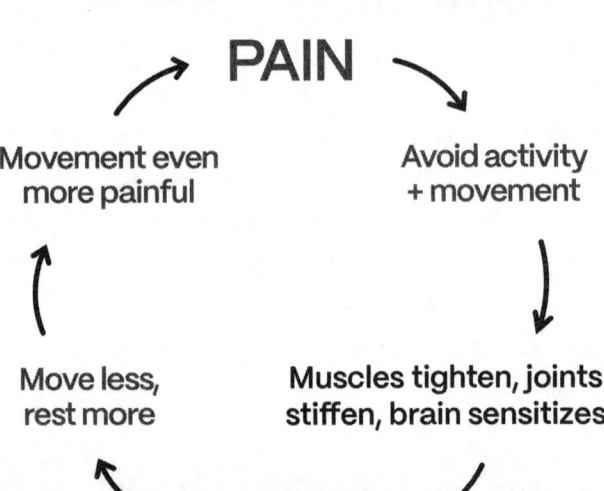

Pain can trap us in an endless cycle of avoidance, inactivity, and more pain.
Credit: Illustration by Eva Huzella

The Boom-Bust Cycle

Conversely, some of us deal with pain by powering through it. We maintain the same intense level of activity regardless of whether pain volume is high or low. Even those of us who prioritize rest on bad days might try to "make up for lost time" on good days. This might look like catching up on work, playing with our kids, exercising, attacking the neglected pile of bills, and reconnecting with friends. But this approach isn't ideal, either: There can be severe consequences for too much work and play. After a boom of activity, we may experience pain, fatigue, and find ourselves unable to get out of bed for days or weeks on end. Ultimately, this boom-bust cycle of intense overactivity and underactivity actually *decreases* functioning over time.

However, there is a solution for this predicament. In order to access it, we must begin by working backward.

THE BOOM-BUST CYCLE

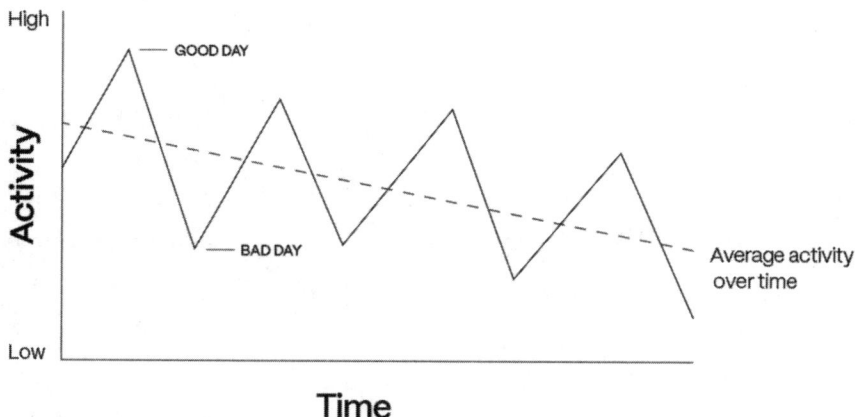

A boom-bust cycle of overactivity and underactivity decreases functioning over time.
Credit: Illustration by Eva Huzella

B. Working Backward: The Power of Pacing

We've been trained to believe that pain needs to resolve before we can resume activities: pain reduction first, activity second. While this is indeed true when it comes to acute pain, like the pain of illnesses, burns, and broken bones, treating chronic pain instead requires a counterintuitive approach called "working backward." This entails gradually resuming activity *first*, exposing our brains and bodies to small, carefully calibrated amounts of stimulation. These intentional doses of activity and movement are tools that create new healing pathways in the brain, strengthen our bodies, and desensitize our overprotective pain system. It is only then that the brain can start unlearning the emergency response to harmless stimuli, like walking or dancing, that might make us hurt but don't cause damage. This requires constantly pushing the "off button" on the pain alarm, reminding our brain that our body is safe—despite the very real pain we feel. And the best way to do this is using a strategy called pacing.

Pacing for pain is much like pacing for a marathon. We wouldn't just hop out of bed and run 26 miles on the day of the race—our bodies would shut down. Instead, we train by slowly increasing activity over time, providing a gentle on-ramp for brain and body to gradually adapt and strengthen. In fact, once we start pacing and resuming activity—seeing friends, getting more sunlight, reengaging in hobbies—our mood is more likely to improve, stress and anxiety more likely to lessen, our bodies more likely to strengthen, and pain volume more likely to dampen. Over time, pacing helps us get better at tolerating whatever activities we choose, even on days when we have some pain. In this way, pacing can result in a gradual *increase* in activity, mobility, and functioning.

Note that a pacing protocol isn't the same as "pushing through the pain." Rather, this method of graded activity helps us identify a safe, comfortable starting point—even if it's just sixty seconds of activity. It also incorporates breaks to rest and stretch as needed. In this way, pacing helps us achieve a balance between doing too little activity, and trying to do too much.

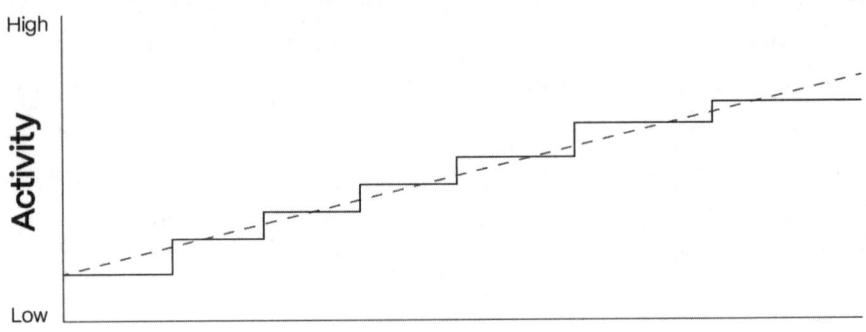

Pacing leads to more activity and functionality over time.
Credit: Illustration by Eva Huzella

ACTIVITY: Crafting Your Pacing Protocol

Pain doesn't discriminate: It affects us at every level of activity and functioning. Whether your goal is to complete activities of daily living, like washing the dishes; to resume beloved hobbies like gardening; or to compete in another marathon, all starting points count—and the same science applies. I've witnessed the most incredible transformations at all levels of ability and disability. After four long years of pain and immobility, pacing helped Sam, the teenager you met in Chapter 1, return to school, socialize, and even play soccer. It helped Sarah the biologist resume using public transit without pain or fear. It helped Joyce the chef start swimming and baking again. These six steps will help you return to life and functioning, too.

Step 1. Set a goal by selecting an activity you'd like to resume, preferably something you miss and used to enjoy. It can be a hobby you've ceded to pain, like playing the piano; a physical activity like walking; or a task you're longing to complete. As long as it's personally meaningful, like soccer was to Sam, it's a great place to start.

Step 2. Break this big goal into a series of small, measurable, achievable steps. Ensure these steps are specific, noting where, when, how, and with whom you'll complete them. Start as small as you need, even if it's just 30 seconds of activity each day. When Sam decided to resume soccer, for example, his first step was to start walking outside daily for one week. He scheduled this walk at 10 a.m. every morning, inviting his sister to join him for company and accountability.

Step 3. Measure the length of time (in minutes) you can comfortably do this activity on a low-pain day, and then on a high-pain day. For example, on a good day, Sam was able to walk outside for about 12 minutes without issue. On a tough day, Sam wasn't able to get out of bed at all (0 minutes).

Step 4. Calculate the average amount of daily activity you can tolerate by adding these times together, then dividing the total by two. For Sam, the equation looked like this: 12 minutes of walking on a good day (low pain) + 0 minutes of walking on a tough day (high pain) = 12 minutes (total). To get the average, he divided the total by two: 12 ÷ 2 = 6 minutes (average amount of daily activity).

Step 5. Once you have an average amount of time you can do this activity, subtract a few minutes to give yourself a cushion. This will be your *baseline*, or your *daily activity goal*—the amount of activity you'll aim to engage in whether it's a high-pain day or a low-pain day. Tempting as it may be, don't do more activity even if you feel great, and try not to do less. Sam's baseline calculation looked like this: 6 minutes (average activity) − 2 minutes (cushion) = 4 minutes = *baseline*, Sam's daily activity goal for week 1. Sam thus walked outside for 4 minutes every day in week 1, whether it was a good day or a bad one, taking breaks to rest and drink water as needed.

Step 6. Each week, add a few minutes to your baseline until you reach your ultimate goal. Some may be able to add 10 minutes, while some will add 2. No shame in this game. Creating an individualized, graded plan that feels good to you is extremely important. For example, Sam's initial baseline was 4 minutes. The second week, he added 3 minutes to his activity, aiming for 7 minutes of daily walking. The following week, his goal was 10 minutes of activity each day. This gradual increase worked for Sam, but a different schedule may work for you. Make sure to build in breaks to rest, stop, stretch, drink water, and pet the dog as needed.

II. Movement Is Medicine

When we live with pain, movement can hurt, let alone vigorous exercise. But because of the brain's capacity to change, physical activity—even in small amounts—can rewire our brain, trigger the release of pain-relieving neurochemicals, and lower the Pain Dial. Indeed, exercise has been shown to do wonders for pain relief, so much so that the phenomenon has a name: exercise-induced analgesia.[3] This is one of the causes of a "runner's high"—a feeling of euphoria that can occur after running. The scientific benefits of movement include:

- Strengthening the body
- Stimulating muscle, tissue, and bone repair
- Facilitating circulation and blood flow for faster healing
- Lubricating joints
- Increasing strength, mobility, and flexibility
- Boosting immune functioning
- Reducing the risk of developing chronic illness
- Improving sleep and reducing fatigue
- Increasing energy and motivation
- Activating mood-boosting neurotransmitters
- Reducing and preventing anxiety and depression
- Eating up stress hormones like adrenaline and cortisol
- Stimulating production of pain-relieving neurochemicals
- Desensitizing the pain system
- Lowering pain volume

For these reasons, movement and activity are critical parts of a chronic pain treatment protocol. Research shows that walking just five to ten minutes a day can have a significant impact on pain and health by reducing inflammation and swelling, and liberating pain-relieving neurochemicals.[4] Walking is so important that many pain management programs recommend making it the first pacing goal. For those who can't afford expensive recovery programs, walking outside is free. And for those with disabilities who are unable to walk, scientists have found that simply peddling our legs in a cycling motion while lying in bed can improve our functioning, increase our chances of recovery, even shorten hospital stays.[5]

ACTIVITY: Movement Medicine

Moving when we live with pain can be hard. The following steps are designed to help you identify manageable forms of movement that can help you heal.

Step 1. Assess your current level of activity. If inactivity or sedentariness have become a part of your life since pain onset, add these to your high-pain recipe on page 180. Then, answer the following questions in your notebook.

- *How many minutes each day do you spend moving and exercising?*
- *What physical activities have you stopped or limited because of pain?*
- *Which do you miss and wish to resume?*
- *Do you have the ability to move any body part for any amount of time?*

Step 2. Below are some good options for moving your body when you live with pain. Consider which might work for you, and add your own:

- Take a walk or go for a run
- Stretch
- Swim
- Water aerobics
- Dance

- Hike in nature
- Lift weights
- Cycle using a traditional, stationary, or recumbent bicycle
- Yoga
- Sex (yes, really!)

Step 3. Aim for one physical activity per day minimum, even if it's just a 5-minute walk. Decide how, when, where, and with whom you'll engage in your selected activities this week. Then enter this date and time into your calendar. This will help you track your goals and maintain accountability. If pain is preventing movement or exercise, use the pacing activity above to help you get out of bed and out the door. Make scheduling daily movement a practice. One good way of doing this is to sit down every Sunday night and map out your movement plan for the week.

PRO TIPS

- If movement sounds daunting or you're unsure how to begin this protocol, start with walking. Use your pacing formula to calculate a safe, comfortable starting point. On that walk, smile at just one person and pet just one dog. Notice how that short walk makes you feel both physically and emotionally.
- *Don't go it alone.* Recovering from pain should never be a solitary activity. There are countless trained professionals, particularly physical therapists and occupational therapists, ready to help us improve balance, strength, and functioning, and establish a successful pacing plan. Social support can come in other forms, too, like a friend or family member. Asking someone to pace or move with us, as Sam did with his sister, can also make achieving our goals more tolerable and enjoyable.

- As much as possible, schedule movement and exercise outdoors to maximize sunlight, fresh air, and social interaction.
- Consider replacing the word "exercise" with the concept of "joyful movement." There are a million ways to move our bodies joyfully. For Kamal, who lived with rheumatoid arthritis, it was playing fetch with his dog, Bear. For you, it might be Tai Chi. What does joyful movement look like for you?
- Lest this seem like an exclusive club that only the able-bodied can join, joyful movement is available to every single person with the ability to move *any* body part. Millie, for example, age seventy-seven and a wheelchair user since childhood, found joy in chair dancing, which involved blasting music—Taylor Swift was her favorite, she told me with an expansive laugh—while energetically moving her torso, arms, hands, and fingers. For our pacing plan, Millie started with gentle movements in her wheelchair for a few minutes at a time. She eventually worked her way up to full-on chair-dancing parties with her family, who joined her every Sunday for a full hour of joyful movement at their local gospel church.
- Healthcare providers: One surefire way to establish an effective treatment plan is to coordinate care across disciplines. A successful approach to pain involves medical doctors, physical and occupational therapists, psychotherapists, social workers, biofeedback providers, and so on. As much as possible, consult with your multidisciplinary colleagues. This is the best way to bridge and reconnect our siloed, disconnected disciplines, and provide the best possible patient care.

Rest and Respite

Just as movement and activity are critical for recovery, so, too, is rest.

In a fast-paced culture that rewards hustle and productivity, the importance of rest is often overlooked. But rest and respite are fundamental parts of any good pacing protocol, and are critical for recovery. Biologically, rest helps the brain and body recuperate, facilitating healing, preventing injuries, and improving our overall functioning. It speeds muscle repair, reduces inflammation, and supports a healthy immune system.

Rest can take many forms. It can mean taking breaks from work by going for a short walk, a weekend of self-care and pampering, or a long vacation. It can mean prioritizing a good night's sleep, or taking a nap (much more on this to come!). Rest plays a particularly important role as you pace to increase activity and movement. Make sure to schedule breaks to rest—to stretch, drink water, take deep breaths, listen to music—before pain gets too intense, and to help you weather flares. Changing position and posture while trying to exercise or increase movement can also help your body reset, and actually increase your tolerance to movement.

Our emotions also benefit from rest. Have you ever noticed that you have more patience and bandwidth—for your kids, work colleagues, even your body—after you've taken some time to rest? Taking breaks regulates our emotions, increasing our ability to tolerate stress, frustration, and pain. Rest can also be cognitive, for your brain. One of the ultimate forms of cognitive rest is silence. Quiet gives our brains a much-needed respite from the constant bombardment of input and noise. It's only when we pause external stimuli—phone calls, podcasts, traffic—and let our minds rest that cognitive processes like introspection, reflection, future planning, and daydreaming are most active.

There is no one-size-fits-all; whatever works for you is the right kind of rest. Ultimately, increasing activity and prioritizing rest aren't mutually exclusive. They can, and should, be done together.

III. The Science of Better Sleep

Anyone who has been in pain knows that when pain flares, sleep suffers. It turns out that the reverse is true, too—when sleep suffers, pain worsens. This cycle is notorious in the world of chronic pain, and it goes something like this:

- → Pain disrupts sleep, causing insomnia, nighttime awakenings, irregular sleep patterns, and poor sleep quality. →
- → Poor sleep leads to exhaustion, fatigue, poorer functioning at work and at home, and attempts to catch up by napping and sleeping in. →
- → Daytime sleep makes it harder to sleep at night. →
- → Lying in bed awake triggers frustration and anxiety, making it even *harder* to sleep. →
- → The worse sleep gets, the higher pain volume climbs.[6] →
- → More pain means even worse sleep. →
- → And 'round and 'round the cycle goes.

Sleep is an important part of maintaining homeostasis and good health. This is particularly true for a body in pain trying to heal. When we sleep, cells repair and regenerate, tissues heal, systems soothe, and new neural pathways grow. But while pain can disrupt these processes, I have two very important pieces of news:

1. *Our bodies are designed to compensate for lost sleep.* Have you ever noticed how, after taking care of the baby until dawn or pulling an all-nighter, your body knocks you out for a nap the next day and sends you to bed early, barely able to keep your eyes open? Evolution has accounted for the possibility that we may not always sleep well, and designed a backup system: Our brains come with a built-in mechanism to compensate for sleep deprivation. If we miss too much rapid eye movement (REM) sleep, for example—the kind of sleep critical for memory and learning—research shows that the brain makes up for it by

increasing the amount of REM we get during subsequent sleep cycles. This phenomenon is known as "REM rebound," and it's a perfect example of homeostasis at work.

2. *Sleep cycles can become disrupted, but they can be fixed.* There are highly effective strategies we can use to get our brains and bodies back on track, and even make up the sleep we've lost. The best of these is a technique called "sleep hygiene."

Sleep Hygiene

As children, we learn the basics of oral hygiene to ensure our teeth don't rot and fall out: Brush twice a day, floss between meals, and visit the dentist regularly. Just as there is a formula for good oral hygiene, so, too, is there a formula for better sleep: a set of scientific guidelines called sleep hygiene that we can use to improve sleep quality and quantity. I know it works because I've been using it and recommending it to patients for more than twenty-five years. If you suffer from sleep issues like insomnia—or "painsomnia," as it's known in the pain world—this can be an incredibly powerful tool. So, what evidence do we have that it actually works?

Research suggests that sleep hygiene can be as effective as sleep medications in the short term, and even *more* effective than medications over the long term—particularly when packaged with a treatment called cognitive behavioral therapy for insomnia, or CBT-I.[7] This treatment protocol targets the unhelpful behaviors, emotions, and thoughts that contribute to sleep issues. By helping to reprogram the brain, sleep hygiene and CBT-I readjust sleep hormones and neurotransmitters, addressing fatigue, skewed sleep schedules, sleep-related stress, and other symptoms.

When these protocols were specifically evaluated for addressing chronic pain, CBT-I and sleep hygiene were found to treat not just sleep issues—they also helped reduce pain. This is because sleep and pain are so intimately interconnected that improving one typically adjusts the other. In fact, one recent study revealed that the probability of having better sleep after implementing CBT-I and sleep hygiene was higher than 80 percent, while the probability of having less pain was nearly 60 percent.[8] These are odds worth betting on, inspiring experts to refer to CBT-I as a "potential pain killer."[9] Ultimately, this evidence reminds us that pain treatments focusing only on a body part—and neglecting other ingredients, like sleep—are dangerously incomplete.

ACTIVITY: Sleep Hygiene Protocol

If poor sleep is an ingredient in your high-pain recipe, add "poor or insufficient sleep" to your high-pain recipe on page 180. The sleep hygiene protocol below will help. It highlights my top ten favorite guidelines, along with the science behind them. Implementing all of these together is the best way to go. However, if this feels too overwhelming, start by implementing just two guidelines, and fold in two more each subsequent week.

1. Get sunlight first thing in the morning. Our brains have a built-in biological clock called the suprachiasmatic nucleus (SCN). This small structure regulates our circadian rhythms—the twenty-four-hour cycles constantly running in the background, carrying out essential processes like hormone release, digestion, and sleep. These daily rhythms, punctuated by sunrise and sunset, determine when we feel alert, when we're hungry, and when we need to stop and rest. Circadian rhythms are the reason bears know when it's time to hibernate and Canada geese know when it's time to fly south. Because the SCN is programmed by sunlight, we have the ability to set and reset our biological clock much like we set our alarm clocks. Getting sunlight first thing in the morning by opening the blinds or stepping outside liberates "wake" chemicals, and breaks down "sleep" chemicals like melatonin, syncing our brain with the sun and the outside world.

2. Set a sleep time and a wake time, and stick to them. Have you ever noticed that if you set your alarm consistently for 6 a.m., your brain will eventually start waking you at 6 a.m. without any help? Brains love routine, because routines are at the heart of homeostasis. Establishing a healthy sleep routine is particularly important when we have pain, a time when irregular schedules can become the norm. Plan to go to bed at approximately the same time every night, and wake at the same time every morning. Consistent sleep times help set our biological clock and regulate sleep. And the earlier you wake up, the earlier you'll be able to fall asleep.

3. Restrict sleep to nighttime: limit napping and sleeping in. After a night of insufficient rest, napping and sleeping late seem the simplest fix. But this is a trap when trying to improve sleep. Over the course of the day, a sleep chemical called adenosine builds up in our brain. Increasing levels of adenosine increases "sleep pressure," or the urge to sleep. When adenosine and other chemicals reach a critical threshold, we fall asleep. Napping artificially reduces adenosine levels and sleep pressure, such that we don't become sleepy until later at night. This dysregulates our sleep cycle and triggers nighttime sleeplessness. Counterintuitive as it may seem, resist the urge to nap if you're trying to improve sleep. If you must nap, limit it to fifteen or twenty minutes, and stick to naps earlier in the day rather than later.

4. Use your bed only for sleep and sex. Our brains are built to make associations between things, related or not. A famous experiment by a Russian scientist named Pavlov demonstrated how this works. By training dogs to expect meat every time he rang a bell, ringing the bell alone eventually incited the dogs to drool—even though bells are neither edible nor delicious. This learned response is called conditioning, and we can use it to help us sleep.

Lying in bed awake and doomscrolling, for example, a common bedtime activity, trains your brain to associate "bed" with "awake and stressed"—the opposite of what we want. The most effective way to program your brain to associate "bed" with "sleep" is to use your bed *exclusively* for sleep and sex, and nothing else. The more you're in bed fast asleep, the stronger this association will become. Find another place to watch TV, scroll social media, and read.

5. *Get out of bed after about twenty minutes when you can't sleep.* The longer we lie in bed awake, in pain and worried about not sleeping, the more we train our brains to associate "bed" with "stressed wakefulness." To ensure our bed is primarily associated with relaxed, restful sleepiness, get out of bed after approximately twenty minutes. (No need to look at a clock; you'll know when enough time has passed.) Plan for this by keeping relaxing activities next to your couch. You can read a magazine, listen to soothing music, draw, or try guided meditations. *Do not* look at your phone, check email, or watch the news. Nothing stimulating or exciting should happen when you can't sleep. Then, when you start feeling sleepy, climb back into bed. Repeat as many times as needed.

6. *Cover your clocks.* We've all checked the time on nights we can't sleep, wondering how many hours we have left until we need to be awake. This instantly increases stress and worry, making it even *harder* to fall asleep. During this period of resetting your biological clock for better sleep, cover all visible clocks—the oven clock, the hall clock, even the clock on your phone. Lay them facedown or cover them with tape or anything handy. As far as we're concerned, time is irrelevant.

7. *Dim your lights in the evening.* Melatonin, also known as the "dark hormone," is a powerful sleep chemical made by the brain's pineal gland (once believed to be the seat of our soul). In addition

to making us sleepy, melatonin also has antioxidant properties that help tissues heal. Dark stimulates melatonin production, and the darker our ambient environment gets, the more melatonin we produce. The higher our melatonin levels, the sleepier we become. Bright light suppresses melatonin production. So, in the evenings, dim your lights, shut off bright overhead lights, and limit screen time.

8. Establish a relaxing bedtime routine. An evening routine cues the brain that it's time to slow down and check out for the day. Your routine can include a warm bath, reading, meditating, stretching—anything soothing and quiet. The more regular your routine becomes, the more likely your brain is to respond. Screens should not be part of this routine, as the blue light that emanates from them artificially cues the SCN that it's time for wakefulness. Screens also activate our sympathetic stress response rather than quieting it. Put screens away a few hours before bed, and replace them with relaxing activities instead.

9. Create a comfortable sleep environment. Our bodies sleep best in environments that are dark, quiet, and cool. Dark and quiet seem obvious. But why cool? Our core body temperature naturally drops at night, and starts to rise again a few hours before we wake. This natural craving for coolness during sleep is why it's harder to fall asleep on hot summer nights and after vigorous exercise. When our environment is too hot—or, conversely, too cold—it can interfere with our internal body temperature, making it difficult to fall and stay asleep. To create an ideal sleep environment, use curtains to block out bright light at night, a white noise machine or earplugs to dim loud noise, and keep your bedroom a cool, comfortable temperature.

10. Limit caffeine, alcohol, and nicotine. Substances dysregulate our circadian rhythms, interrupting our sleep-wake cycles and preventing us from tuning in to natural sleep signals. Caffeine and nicotine artificially speed us up, while alcohol artificially slows us down, making it harder to get back on schedule. These substances also disrupt sleep, leading to insomnia, nighttime awakenings, and poor sleep quality. This doesn't mean we shouldn't enjoy coffee in the mornings or wine with dinner. However, during periods of pain and poor sleep, cutting back can be medicinal.

PRO TIPS

- Like all habits, sleep habits can be hard to change. The more we stick to our new sleep hygiene protocol, the easier it will become to implement, and the better our results will be.
- Sleep medications aren't recommended for chronic sleep issues, particularly those associated with chronic pain. Here's why:

1. Sleep medications mask symptoms rather than addressing underlying root causes, like pain or anxiety.
2. The chemicals found in sleep medications *disrupt* circadian rhythms and sleep cycles rather than supporting them.
3. Sleep medications are designed for short-term use. People with chronic sleep issues are better served by strategies designed for long-term use, like sleep hygiene and CBT-I.
4. Sleep medications usually don't mix well with pain medications, and can be dangerous—even lethal—when combined.
5. Many sleep medications have addictive properties, leading to an overreliance on them to fall asleep. Adding this burden on top of pain is the last thing you need.

- If you've been taking sleep medications for a long time, particularly benzodiazepines, don't suddenly stop using them, as this can be very dangerous. Make sure to speak with your doctor to establish a safe tapering plan if you want to reduce your use.
- To find a trained CBT-I provider, try the CBT-I directory (www.cbti.directory) or the Society of Behavioral Sleep Medicine (behavioralsleep.org). If you can't find a CBT-I provider near you, any trained cognitive behavioral therapist can help you implement a sleep hygiene protocol. Free sleep hygiene protocols are also available in *The Pain Management Workbook* and online.
- Sleep issues can be signs of other health conditions, too, like sleep apnea. These require a medical assessment for diagnosis and appropriate treatment.

IV. Nutrition and Pain: Eat Better to Feel Better

While food may seem to be a negligible part of treatment, the foods we consume not only provide us with energy and fuel, like gas in a car—they also become the building blocks of our musculoskeletal, circulatory, immune, and nervous systems. These systems play critical roles in healing. And when they break down, so does the healing process. Our diet also affects our neurochemistry and hormones, impacting everything from sleep to our emotional health. How much we eat and when affects our mood and health, too. We've all experienced the relaxed bliss of "food comas," like the kind induced by a Thanksgiving meal, or, conversely, have become "hangry"—so hungry that we become irritable and angry.

While certain foods, like fruits and vegetables, are packed with vitamins and minerals that promote health and healing, others—like processed and artificial foods—are full of preservatives and chemicals, and can have deleterious health effects. Unfortunately, the standard Western diet is packed with foods and ingredients known to trigger inflammation and amplify pain, like processed foods, saturated and trans fats, chemical additives and preservatives, and excess sweets and treats.[10]

Poor nutrition turned out to be a high-pain ingredient in Sam's pain recipe. Early in his treatment, I learned that Sam was on a largely "white food" diet, one consisting of bread, pasta, cereal, and pizza. His parents couldn't get him to eat fruits or vegetables, they said, and they didn't want to push him given how miserable he already was. Thus, one of my first interventions was helping Sam improve his diet, giving him the energy he needed to control his Pain Dial and take back his life. But this pattern isn't unique to Sam. For many of us, nutritional inadequacies quietly fuel the Pain Cycle.

This is in part because chronic pain is a risk factor for eating poorly. When we have less ability and energy to shop, prep, and cook, we're more likely to reach for the simplest option—whether that be fast food or a bag of chips. Others skip meals, eat too much or too little, or develop unhealthy eating habits, such as eating primarily "comfort foods" like fried chicken and pizza. But while food can be soothing, an imbalanced, insufficient diet disrupts homeostasis, triggers fatigue and brain fog, and perpetuates and amplifies pain.[11] Vitamin and mineral deficiencies also prevent our immune system from functioning properly, increasing our susceptibility to illness and making it harder for our bodies to heal. In this way, inadequate nutrition not only can make us sick, but *it also prevents us from getting well.*

Fortunately, the opposite is also true. When scientists reviewed forty-three clinical studies in which people living with chronic pain were prescribed a diet packed with healthy, whole foods, the results were undeniable. Compared to control groups, those provided with more nutritious diets experienced clinically meaningful pain reductions—a 33 percent reduction in overall pain!—compared to those who didn't.[12] This has serious implications for people living with pain. As part of our pain protocol, a quality, nutrient-rich diet is an important low-pain ingredient that can adjust our dial. This means upping our intake of fresh, whole foods and antioxidants, including fruits, vegetables, whole grains, dairy products, lean proteins, and healthy fats. The following are foods that can optimize our ability to fight pain and promote healing:

Low-Pain Foods
- Fruits: cherries, red grapes, blueberries, strawberries, apples, oranges, pears, bananas
- Vegetables: dark leafy greens, carrots, beets, broccoli, brussels sprouts, cabbage, cauliflower, kale
- Whole grains: brown rice, lentils, oatmeal, quinoa, couscous, whole wheat bread, whole grain pasta, popcorn
- Eggs
- Dairy: milk, yogurt, cheese
- Fish
- Black beans
- Nuts and seeds
- Teas containing antioxidants (e.g., chamomile tea, green tea)
- Ginger
- Olive oil (in place of butter, margarine, or other substitutes)

Simultaneously, we want to reduce our intake of foods that can trigger inflammation or amplify pain. These include caffeine, which disrupts sleep and activates our stress response, and alcohol, which interrupts circadian rhythms and weakens our immune system over time. In choosing the best fuel for a body in pain, we also want to minimize saturated and trans fats; processed foods with ingredients we can't pronounce; diet foods and drinks full of chemicals; and excess sugar, which goes by many names, often ending in "ose," like sucrose and fructose. Ultimately, what we put into our bodies determines what we get out of them.

High-Pain Foods
- Fast food
- Fried foods
- Processed foods (e.g., chips, breakfast cereals, frozen pizza, processed meats like hot dogs and lunch meat)
- Foods high in saturated and trans fats
- High-sugar foods like soda, candy, cookies, and sweets
- Caffeine (e.g., energy drinks, coffee, yerba mate)
- Alcohol
- Diet drinks

Improving nutrition to reduce pain need not mean a dramatic shift in how we eat, or giving up our favorite foods. Dietary changes are most successful when they're small, realistic, measurable, and achievable. Pacing doesn't just apply to movement; it can also support our nutritional goals. Simply adding one fruit or vegetable to each meal while cutting back, even moderately, on fried foods, diet drinks, or sugar is a great place to start. For example, plan to stack your lunch sandwich with lettuce, spinach, and tomatoes a few times a week. Snack on apples and carrots in the afternoons. Replace french fries at dinner with lentils or a sweet potato. Consider reducing alcohol intake by just one drink per week. Upping our nutrition game also means tending to how often we eat, and how much (or little!), ensuring that we get three healthy meals a day plus snacks rather than skipping meals or settling for a bag of chips.

ACTIVITY: Nutrition Protocol for Pain

Step 1. Conduct a simple self-assessment of your nutrition and daily diet, logging the answers to the following questions in your notebook.

- *Do you eat three healthy, balanced meals each day? If not, when are you most likely to eat something processed or unhealthy, or opt for fast food? When are you most likely to eat beyond full? If you skip meals due to pain, which meals are you most likely to skip?*
- *Do you eat fruits and vegetables with every meal? If not, what gets in your way?*
- *How much caffeine and alcohol do you consume? What do you notice about how these affect your sleep, pain, or health?*
- *Have you been fasting or crash dieting?*

Now review your answers to the questions above. Which eating habits, dietary insufficiencies, or unhealthy foods do you suspect might be triggering, prolonging, and maintaining your pain? Add these to your high-pain recipe. Be specific, including details like "eating high-pain foods," "crash dieting," or "insufficient nutrition."

Step 2. Review the list of low-pain and high-pain foods in this section. How can you integrate more low-pain foods into your daily intake, and/or swap less healthy options for these healthier ones? Make a list of three low-pain foods you're willing to increase this week, and three high-pain foods you're willing to decrease.

Step 3. Troubleshoot obstacles in advance. Common obstacles to changing our diet include the cost of groceries, access to healthy foods, prep time, motivation, the challenge of overcoming old habits, and being conditioned to view unhealthy foods as rewards. Identify your obstacles, and generate some ideas for troubleshooting them. For instance, if prep time is an obstacle, make a plan to prepare and freeze healthy meals for the week on Sunday. If you've been using unhealthy foods as rewards, brainstorm nonfood rewards you can use in its place (a massage, a concert, a trip to the museum, time with friends).

Step 4. With these in place, craft your nutrition protocol—a personalized plan for upping your intake of low-pain foods, and reducing high-pain foods, every day of the week. It should look something like this:

DAY OF THE WEEK: *MONDAY*

Breakfast	Add one piece of fruit (orange, banana)
Lunch	Switch to whole grain sandwich bread
Snack	Bring an apple and nuts to the office instead of hitting the vending machine
Dinner	Add a vegetable side dish, substitute butter with olive oil when cooking

PRO TIPS

- The goal here isn't to "go on a diet," lose weight, count calories, or conform to a particular body shape. Restrictive diets don't work anyway, resulting in yo-yo weight loss and gain, and inducing binging behavior when we feel deprived. They are also *bad* for pain. Rather, this is an invitation to continue eating your favorite foods while gradually replacing inflammatory foods with healthier options that optimize your ability to heal.
- For the purposes of pain management, reject fad diets like the ketogenic, paleo, and carnivore diets, as well as

intermittent fasting. While these might serve other purposes, they provide insufficient nutrition for healing—and there's zero evidence they reduce pain. In fact, fasting is a known pain *trigger*, causing headaches, muscle aches, stomach cramps, and discomfort due to low blood sugar and dehydration. Some studies suggest that fasting can even reduce our ability to tolerate pain, making whatever pain we have feel worse.[13] If you have other medical conditions that require dietary constraints, like diabetes or allergies, always stick to the nutrition plan recommended by your doctors.
- Supplements aren't a panacea and can even be harmful. There isn't a single supplement on earth that magically cures pain. In fact, there's woefully little evidence that supplements are actually useful unless you truly need one.[14] A "supplement" is so named because it supplements any deficiencies you might have. If you're deficient in vitamin D, for example, you might benefit from a vitamin D supplement (and/or more sunlight!). The best way to determine whether your body actually needs a supplement is to ask your doctor to conduct a blood panel and other lab tests. These tests can identify deficiencies, which can usually be easily corrected with dietary changes. Don't waste your hard-earned money on supplements unless your doctor has actually advised you to use them.
- There is no "superfood" that cures pain. Ignore social media fads.
- Changing the way we eat requires making healthy decisions at every meal, which can be harder than it sounds. If you've historically struggled with food, weight, or diets, a consultation with a registered dietitian can provide support, guidance, and accountability.

The Miracle of Pain Medications

We can't conclude the "bio" chapter without addressing pain medications. While reviewing all available medications is beyond the scope of this book, I want to be extremely clear: *Medications are absolutely appropriate for pain.* Pain medications are a miracle of modern science. Without them, we'd still be biting on wooden spoons during surgery. Opioids and other anesthetics, anti-inflammatories, and other medications are a godsend for amputations, childbirth, root canals, and other painful procedures. Over-the-counter drugs are a gift when managing injuries, headaches, and the acute pain flares of chronic conditions like cancer and sickle cell anemia.

However, while we shouldn't be afraid to use pain medications, particularly for acute pain, they are not a cure for chronic pain. We all wish they could be. But medications are only a Band-Aid. They do not change our brains and bodies in a way that reduces pain over the long term. In some cases, pain medications like opioids can even do the opposite—making our pain system *more* sensitive instead of less.[i] This means that, when we try to go off them, we don't feel better—instead, we feel worse. Over-the-counter pain medications have limitations, too: They're designed to be taken only a few days in a row, at the specific dose marked on the label. Taking too many for too long can cause health issues like stomach bleeding, liver damage, kidney failure, and worse.

Moreover, the human body builds a tolerance to pain medications over time, meaning that it takes increasingly higher doses to achieve the same effect. And that's to say nothing of their side effects. Steroids, for example, commonly prescribed for joint and musculoskeletal pain, can actually harm our bodies over the long term, slowing wound healing, thinning our bones, and even compromising our immune system.

Relying on drugs to regulate pain volume also cultivates a sense of

---

i   This is particularly true of opioids, which sensitize the brain to pain and ultimately make pain feel worse over time. This is called "opioid-induced hyperalgesia," or increased pain caused by long-term opioid use.

powerlessness—a false perception that we need to reach outward instead of inward to control our Pain Dial. All told, for chronic pain, there's no evidence to suggest that pain medications are the answer—in contrast to abundant evidence that they can cause more problems than they solve.[15,16] This is true of surgeries, too: up to 60 percent of adults (and 20 percent of children) *develop* chronic pain after surgery.[17] It's no surprise, then, that experts around the globe agree that neither medications nor surgeries are the best, safest, or most effective solution for chronic pain. All of my patients, everyone you've met here, had to learn that the hard way. Perhaps you have, too.

So, buyer beware: Anyone selling a magical cure for pain, whether it be a newfangled gadget, medical device, or "plant-based medicine" as a quick fix is selling snake oil. (Just because something is "plant-based" doesn't mean it's good for us. Tobacco is a plant, as is poison ivy. And let's not forget hemlock, the infamous deadly poison used to kill Socrates.)[ii]

This doesn't mean you need to stop taking your pain medications. If they're working for you, keep using them. You may, however, discover that you're able to reduce your reliance on them once you have a full tool kit of effective strategies. On the flip side, if you're taking pain medication and still have pain, consider following my three-month rule: If a drug hasn't meaningfully changed your pain within three months, it's time to talk to your doctor about tapering off and trying another treatment plan.

This is also not a suggestion to deny long-term users pain medications. Given that we typically fail to offer affordable, accessible alternatives for pain relief, to abruptly cut off patients' pain meds would be unethical and cruel. When it's necessary to reduce medications, this should be

---

ii This includes kratom, a recent newcomer on the pain scene. The Food and Drug Administration and other public health agencies warn that kratom is being falsely marketed as a safe treatment for chronic pain, when research is clear about its dangers and limitations. Rather than being helpful or curative, kratom is a potentially addictive substance with stimulant and opioid-like effects that has caused thousands of deaths in recent years, triggering a rash of wrongful-death lawsuits. Source: Kaur, H. (2019). More deaths have been associated with Kratom than previously known, CDC study finds. CNN.

done ethically, gradually, and with simultaneous implementation of effective and affordable tools (like the ones offered in this book). Moreover, there are always exceptions—like cancer pain, for example, which usually requires its own set of rules, as well as medications for end-of-life care.

What I hope to leave you with is simply a call for change—a reminder that while medications are sometimes appropriate, there is no single, standalone treatment for chronic pain. In fact, if your pain management program doesn't offer or recommend some combination of physical therapy, occupational therapy, psychotherapy, trauma treatment, sleep and dietary changes, social support, and medical interventions, it isn't actually a pain management program.

Of all the pills we've swallowed, this one is easily the toughest. But if there's one thing we now know, it's that *a whole-person problem requires a whole-person solution.*

* * *

Targeting sleep, nutrition, movement, and medications is vital to our recovery. However, for most of us, it simply isn't enough. Now that we've spent some time on your physical health, let's look at some strategies to bolster your emotional health—which, as we've seen firsthand, is inextricably connected to our physical health and everything happening inside our bodies. These tools are rooted in science, easy to learn, and can fundamentally alter both our psychology *and* our biology.

12

The Emotional Health Protocol

WHY EQ MATTERS IN MEDICINE

To effectively treat chronic pain, we must address not only our emotional well-being and mental health, but also our EQ: our emotional quotient, or emotional intelligence. Developing our EQ includes learning how to better identify, express, and manage the full spectrum of human emotions. Given how significantly emotions like stress, depression, and optimism change pain volume, this skill is as critical to treating pain as medical know-how. Don't fall for the tired trope that these interventions are "woo" or pseudoscience: They exert their effects by altering our physiological systems, activating the body's pharmacy, and changing our brain pathways.[1,2] While these are not a panacea on their own—no single treatment is—when combined with a complete pain protocol, they pack a great punch.

I. An "Off Button" for the Pain Alarm

Retraining the Pain System: The Role of Relaxation

Every emotion we have changes the pain we feel. This can make it challenging to know where to start. I usually aim for the biggest and most obvious target first: stress. Because if stress is a pain megaphone, we'd all do well to learn how to operate the mute button. There's a reason that top pain management programs recommend interventions like mindfulness and relaxation strategies, and why muscle relaxants are at the top of the pain prescription list.

Our sympathetic nervous system (SNS), which is also our stress-response system, is counterbalanced by the parasympathetic nervous system (PNS)—otherwise known as our rest-and-restore system. Just as a sustained stress response amplifies pain, relaxation, stress reduction, and states of deep rest turn pain volume down. The PNS is one of our body's many built-in healing apparatuses, responsible for facilitating immune function, conserving energy, and speeding tissue repair.

Modern science has developed all sorts of methods for reducing an overactive SNS response while stimulating the PNS. These methods have one thing in common: They communicate to our brain that we are safe, telling it that there is no danger or threat, thus lowering the pain alarm. These strategies are bidirectional: By soothing our brains, we soothe our bodies, and by soothing our bodies, we soothe our brains. Think of these techniques as "off buttons" for the pain alarm—buttons you can press anytime, day or night, as many times as you need.

Identifying Triggers, Stressors, and Amplifiers

Stress is additive, as is its impact on the body. The more stressors—also known as pain triggers—we have, the more they'll amplify our pain. But before we can learn to mitigate their influence, we first need to know what our stressors are. Some are silent; others are sneaky. Investigating your stressors will thus require some detective work. So that tracking them doesn't get unwieldy, try grouping them into categories, as I do here:

- Social stressors: related to family, friends, romantic relationships, or loss/absence of social support
- Occupational: career, employment, school
- Economic: finances, bills, housing
- Environmental: news, politics, social media, pandemic, war

- Major life events or changes: a move, a pregnancy, a scary diagnosis, death of a loved one
- Physiological: pain, injury or illness, mental health issues, hunger or sleep deprivation, being out of balance

ACTIVITY: Identifying Triggers, Stressors, and Amplifiers

Step 1. Some questions to ask yourself as you consider pain triggers and stressors are:

- *How do I know when I'm stressed or worried?*
- *Have I been thinking stressed or anxious thoughts?*
- *What does my body feel like? Is my heart racing, am I talking faster, am I irritable?*
- *What's been on my mind? What keeps me up at night or preoccupies me during the day?*

Step 2. In your notebook or laptop, make a list of your stressors using the categories above. Plan to add additional as you discover them.

Step 3. Then add these to your pain recipe under the "Psych" column on page 178. You can also add them to your list of high-pain ingredients on page 179. These visuals help us track and organize our pain triggers and amplifiers, and point us to the necessary solutions.

Step 4. Brainstorm some ways you can reduce your stressors. While we can't rid ourselves of them entirely, we *can* make efforts to regulate how much—or how little—they impact our health. Over which stressors do you have some control? What steps do you need to take to reduce the impact of these pain triggers and amplifiers? How can you effect change to better protect yourself?

Learning to identify stressors, and brainstorming ways to decrease them, is an important first step toward expanding your EQ and learning to lower pain volume. Once you have a sense of your triggers and amplifiers, you're ready to learn some strategies for regulating your sympathetic and parasympathetic nervous systems. This will give you more control over the impact stress has on your body—and your pain.

Diaphragmatic Breathing

When I teach my patients strategies for controlling the pain alarm, I usually start with breathwork. Simple and free, it has the added benefit of being accessible to you all day every day, no matter where you are.

Breathwork is a technique that involves manipulating and regulating the breath to induce physical and cognitive changes. It's extremely popular these days, used for everything from cold plunges to Olympic training. One of the simplest, most effective forms of breathwork is diaphragmatic breathing, named for the diaphragm, the large abdominal muscle just below our rib cage that expands and contracts as we breathe. Surprisingly, most of us aren't doing the best job of using it. When we're stressed, sick, or in pain, we tend to chest breathe, a shallow breathing pattern that relies on the muscles between our ribs and neck to inflate our lungs. Chest breathing maintains stress and anxiety, provides insufficient oxygen to brain and organs, restricts blood flow, and perpetuates pain system hyperactivity.

In contrast, diaphragmatic breathing engages our diaphragm and abdominal muscles, allowing us to breathe deeper, lower, and slower. This reduces SNS arousal, reverses physical symptoms of stress and anxiety, improves circulation and blood oxygen to facilitate healing, and lowers the pain alarm.[3] This is an example of the influence of emotions on your pain alarm in action. Because while chest breathing sends stress and danger cues to the brain, abdominal breathing signals the opposite, informing our pain system that we are safe and that danger messages aren't required.

Diaphragmatic breathing is easy to learn, requires no gadgets, and takes only a few minutes of your time.

ACTIVITY: Diaphragmatic Breathing

Check in with your body and emotions before and after this activity, rating your pain level and stress/anxiety level in your notebook. You can use the traditional 0 to 10 scale (0 = none, 5 = moderate, 10 = severe) or simply make a note of how you felt physically and emotionally before and after this exercise.

Step 1. Find a quiet place to sit or lie down somewhere you won't be disturbed. Turn off your screens and put them away. Uncross your arms and legs, and close your eyes.

Step 2. Place one hand on your chest and the other on your belly. Take a slow, deep breath in through your nose. Feel the air traveling down into your lungs, then down into your belly. Fill your belly with air as if it's a balloon. Feel the hand on your belly rise as you breathe in. The hand on your chest should not move. If it does, send that air lower, into your belly. Hold it for just a moment, feeling the tension in your abdomen.

Step 3. Slowly breathe out, releasing the air through your mouth. Extend your exhale, slowly expelling all air from your belly. Feel the hand on your belly sink down. Allow all the muscles in your body to let go and relax. Let your shoulders drop and your body get heavy.

Step 4. Continue breathing into your belly for the next five minutes with eyes closed. Let your breath be low and slow. When you're done, wiggle your fingers and toes, and open your eyes.

PRO TIPS

- Check which type of breathing you're doing throughout the day, stressed (chest) or relaxed (abdomen), by placing one hand on your chest. If your hand is moving, this is a sign that you're chest breathing. This check gives you

> the opportunity to change course. When you catch yourself chest breathing, stop, put a hand on your belly, and instead breathe lower and slower. (Of note, we also use our lungs and chest muscles more when we exercise or escape danger—this is adaptive and requires no change.)
> - You may notice some physical sensations at first, including lightheadedness. This is normal. When people first practice diaphragmatic breathing, it's common to "overbreathe"—exaggerating our ins and outs. This skill will become more natural with time.
> - Diaphragmatic breathing works best when it becomes your default mode of breathing—one that you perform automatically, without needing to focus or concentrate. The more you practice, the easier, more automated, and more effective it will become.

II. Biofeedback: Reconnecting Emotions and Sensations

Biofeedback, the technique I first discovered in Dr. Peper's office while incredulously warming my hands—not over a fire, but using my brain—is a tool that allows us to receive feedback about unconscious biological processes, like muscle tension and heart rate, and then change them. In the first phase of biofeedback, sensors affixed to your body gather data about your physiological state—breathing patterns, skin conductance, and other bodily functions. The sensors then feed this information to a display so that you can observe, in real time, the way emotions and thoughts change your body. (You may already be wearing a biofeedback device, by the way: That smartwatch that reads your heart rate and tells you when it's time to stand is one!)

In the second phase of biofeedback, you learn to retrain your brain and body using relaxation strategies, breathwork, and visualization techniques. You then watch the screen as your body responds. When stressed, for example, you can actually observe as muscle tension increases, heart rate spikes, and body temperature drops. Dr. Peper demonstrated this to me in his office, when he had me generate my long to-do list. Then, as he guided me through relaxation, diaphragmatic breathing, and visualization techniques, I was able to watch as my brain learned to reverse it.

There's incredible power in using biofeedback as a pain management tool. When we see these changes occurring before our very eyes, it's visible, tangible proof that our emotions and thoughts are connected to physical sensations—and that we can exert some control. When I teach biofeedback techniques like hand-warming to patients, the power it generates is palpable. I often get responses like "If I can make fireballs with my hands, *what else can I do with my mind?*" This is exactly the kind of agency and hope we want to instill.

A ton of research has been done on biofeedback, and there's no doubt it can reduce pain. It has evidence for ameliorating the duration and intensity of sickle cell flares, cancer and chemotherapy pain, chronic back pain, pelvic pain, fibromyalgia, abdominal pain, IBS, irritable bowel disease, diabetes, hypertension, and anxiety, among other chronic conditions.[4] Biofeedback has proven particularly effective for migraine and headaches, resulting in a 60 percent reduction in headache frequency and severity—equivalent to the effects of popular headache medications, but without the side effects.[i] You can now harness its power, too.

---

i Biofeedback has been shown to be as effective for headaches as popular, commonly prescribed migraine drugs propranolol (Inderal) and amitriptyline (Elavil). Source: *Biofeedback and Relaxation Training for Headaches.* American Migraine Foundation, Nov. 12, 2016. https://americanmigrainefoundation.org/resource-library/biofeedback-and-relaxation-training.

> **PRO TIPS**
>
> - Working with a trained biofeedback provider in person is the optimal way to receive treatment tailored to your unique body and needs. Find a provider near you via the Biofeedback Certification International Alliance, bcia.org.
> - You can also try biofeedback via an app, many of which are low-cost or free, like HeartMath or Breathe2Relax. My patients regularly use biofeedback apps and find them very helpful.
> - Biofeedback is most effective when you practice daily—once a week at a minimum.
> - <u>Healthcare providers</u>: Consider having a trained biofeedback provider on your treatment team or getting certified yourself via bcia.org. One of my former trainees at the Stanford University School of Medicine, an addiction medicine fellow, decided to get certified after learning about biofeedback in our class, and has found it helpful for his patients.

III. The Neuroscience of Mindfulness

A recommendation to try meditation for pain invariably results in an eye roll. "My pain is real. You want me to just *meditate* my pain away?" It's a valid response. At first glance, mindfulness appears to target only the mind. But what does neuroscience say?

Mindfulness can have an enormous impact on our health and our pain. It strengthens connections in the prefrontal cortex, improving our ability to regulate attention and emotion. It reduces amygdala and SNS activity to decrease stress and anxiety, and can improve sleep and mood. It also changes the functioning of our pain system, decreasing activity in brain regions involved in the emotional and cognitive control of pain.[5] Mindfulness is so effective that it can even reduce our need for opioids.

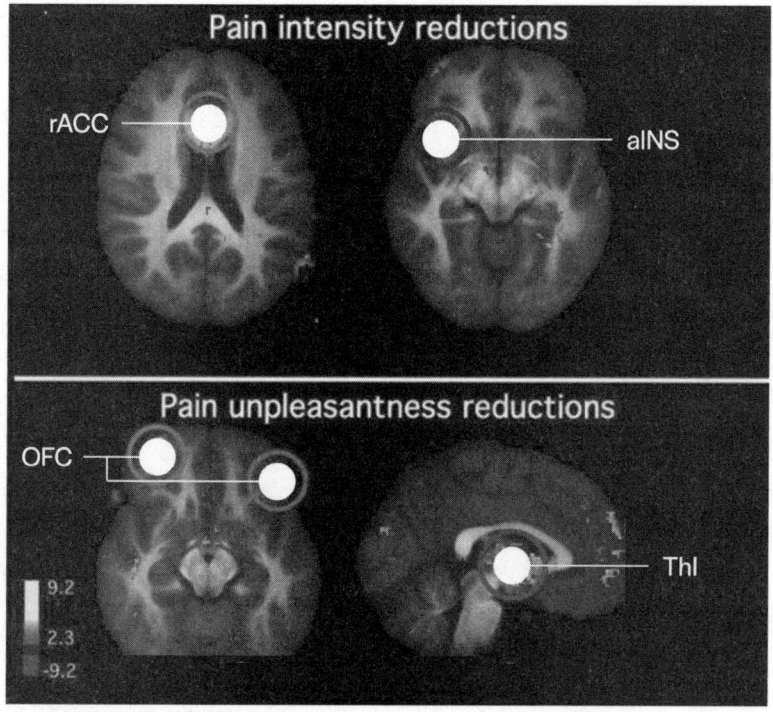

Mindfulness significantly reduces the brain's pain response.
(rACC = right anterior cingulate cortex. aINS = anterior insula.
OFC = orbitofrontal cortex. Thl = thalamus)

Credit: Image adapted from Neuroscience Letters, *volume 520(2), Zeidan F., Grant J., Brown C., et al. Mindfulness meditation–related pain relief: Evidence for unique brain mechanisms in the regulation of pain, pp. 165–173. Copyright © 2012 Elsevier Ireland Ltd., 2012.*

In fact, neuroscientists at UCSD have demonstrated that meditating for just twenty minutes a day for four days can produce significant reductions in pain intensity and unpleasantness.[6] Over time, the benefits only multiply: A regular, ongoing mindfulness practice can help decrease the intensity and frequency of pain flares, reduce disability, even improve our quality of life.

One mindfulness-based approach proven particularly successful for pain management is Mindfulness-Based Stress Reduction (MBSR), a scientifically supported treatment that integrates mindfulness and relaxation to calm the mind, relax the body, change brain activity, and lower pain volume. MBSR is easily accessible, with online courses and workbooks available to patients and providers alike.

So, how does mindfulness work? Simply put, mindfulness is the muscle we use to bring our mind from the past *(Why did I do that? How could she say that? What about those terrible things that happened?)* and the future *(What's going to happen to me? What if I never get better? I'm going to hurt forever)* into the present moment. Just right here. Because chances are, you can handle this moment…and this next one…and this one, too. Being in the here and now calms our brains and bodies, reducing stress and anxiety and lowering the pain alarm.

Mindfulness also helps us cultivate an awareness of, and tolerance to, what's happening around us and inside of us without judgment or criticism. Building this skill helps us bear the unbearable, and tolerate the intolerable. This tool is particularly important for people living with pain—because fighting pain, ignoring it, and pushing it away, while a tempting and natural impulse, only makes pain *worse*. Sometimes we can't disappear pain, and it's just a fact of life. Mindfulness transforms our relationship with pain, allowing us to be with difficult sensations and emotions in new ways.

Mindfulness also teaches us how to insert a pause between trigger and response, which can be a game-changer. Pain automatically triggers our SNS stress response, causing us to clench, brace, and tense. But this only makes pain worse. Mindfulness offers us the opportunity to stop, breathe, center—and purposefully select better, more effective reactions. For example, we might allow pain to move through us while initiating breathwork and relaxing our muscles. This, in turn, activates the parasympathetic rest-and-repair system, and helps lower the pain alarm. Learning to use mindfulness as a tool helped Joyce, dancer and chef. She started noticing when she was tensing her back and shoulders—which, she realized, seemed to happen every time she was in pain, thought about her ex-husband, or approached her endless pile of bills. As soon as she recognized she was tensing, and learned her triggers, she was able to start doing the opposite: releasing and relaxing her shoulders, neck, and back—finally interrupting the cycle of pain.

MINDFULNESS CHANGES THE PAIN RESPONSE

Old Pattern
pain → SNS stress response → tense, brace, clench, get stressed → pain amplification
New Pattern
pain → mindfulness → allow sensations to pass, breathwork, relaxation strategies → muscle relaxation, PNS activation → pain reduction

In this way, mindfulness helps us cultivate four important skills for pain management:

1. *Somatic self-awareness.* Mindfulness helps cultivate increased awareness of, and ability to tolerate, pain and physical sensations. We can better notice, for example, how pain is never fixed—but is instead impermanent and always changing. This is a critical part of recovery, because it reminds us that if pain is always changing, that means pain can change.
2. *Increased emotional awareness.* When we practice mindfulness, we increase our ability to tune in to, notice, and label emotions. We start to recognize when they arise, what triggers them, and where we feel them in our bodies.
3. *Relaxation and stress reduction.* Being in the present moment, rather than focusing on the past or future, is a well-established stress reducer. Becoming more mindful of bodily responses to stress and pain also provides an opportunity to release hidden muscle tension, reduce SNS activity, increase PNS activity, and foster a calm, safe state.
4. *Cognitive control.* Practicing mindfulness cultivates enhanced awareness of mental activity—that is, what our minds are thinking and focusing on in any given moment. This gives us better ability to regulate attention and change negative thought patterns.

ACTIVITY: Body Scan for Pain

The mindfulness practice I teach most, and usually first, is the body scan, which combines progressive muscle relaxation with meditation and stress reduction. One of the reasons we can scan our bodies using our brain is because we have a map of our entire body that lives in our brain, called the homunculus. This neurological map contains data about our body, including how it feels (sensation) and what it's doing (movement). If you sense into your left foot, for example, noting its placement on the floor and how your sock feels on your skin, you're activating your homunculus.

When we practice the body scan, we use our brain map and other sensory powers to investigate our body from head to toe without judging sensations or emotions, or trying to push them away. As you do the body scan, you might discover hidden muscle tension, anxious thoughts, or negative emotions: potent high-pain ingredients that would otherwise sneak by undetected. The body scan also helps us cultivate a state of deep rest, lowering heart rate, blood pressure, and even brain wave frequency. This allows our PNS to do its important self-healing work.

Here's how to do it:

Check in with your body and emotions before and after this activity, rating your pain level and stress/anxiety level in your notebook. You can use the traditional 0 to 10 scale (0 = none, 5 = moderate, 10 = severe), or simply make a note of how you felt physically and emotionally before and after this activity.

Step 1. Find a quiet place somewhere you won't be disturbed. Lie down and turn off your screens. Make sure you're comfortable and warm. Uncross your arms and legs, and close your eyes.

Step 2. Imagine your attention is a spotlight, and you can control where it shines. Shine that spotlight on your breath. Notice the cool air at your nose. Watch the air travel down into your lungs, then down into your belly. When you exhale, notice how the air feels leaving your body. There's no need to try to breathe any particular way. Just let your body breathe on its own.

Step 3. Next, shine that spotlight of attention on your thoughts. Without judging them, take note of what your brain is doing and simply label it: worrying, planning, remembering. Don't try to stop, control, or avoid these thoughts. Just let them come and go, like clouds drifting in the sky.

Step 4. Tune in to your emotions, noticing how you feel. Label each emotion without judgment: stressed, sad, peaceful. Take note of where you feel these emotions in your body: a heaviness behind your eyes, throbbing in your temples, a clenched jaw, tightness in your stomach.

Step 5. Now use that spotlight of attention to slowly scan your entire body. To begin, start by noticing how the couch, bed, or floor is holding you. Notice that you are safe. Send the spotlight to the top of your head, and notice the sensation of your head on the couch. Move that spotlight down slowly: into your forehead, eyes, facial muscles, and neck. Take your time. Scan down into your shoulders, chest, and abdomen. Sense now into your back, down your arms, into your fingers. Slowly scan your hips, legs, and down into your feet. If you discover any muscle tension, see if you can let that tension go, allowing your muscles to soften and relax.

Step 6. Notice, without judgment, all sensations in your body. Describe each one to yourself: tingling, burning, numb. Examine places of pain and discomfort without wishing them away. Breathe into that body part, inviting it to relax on the exhaled breath. Ride the wave of sensations, noticing how they change, shift, and move. Allow all sensations to come and go, noting their impermanence.

Step 7. Take inventory of your entire body now, noting all sensations, emotions, and thoughts. Take a few slow, deep breaths. Place one hand on your heart and wish yourself well with kindness, as if addressing a dear friend. When ready, wiggle your fingers and toes, and open your eyes.

PRO TIPS

- The goal of mindfulness isn't to clear your mind or control your body, but rather to be present with everything happening inside you and around you.
- When we do the body scan for pain, it's helpful to add "safety reminders," noting that your body is safe. This helps lower the pain alarm, and is particularly important if you're feeling stressed, anxious, or have a history of trauma.
- Once you learn how to do it, you can use the body scan anyplace, anytime. Practice in your car, the waiting room at the doctor's office, at work. Because the brain is plastic, the more you practice this skill, the better you'll get at it.
- Try one body scan a day as if taking a daily pain medication. Find a day and time that works for you, and then stick with it until it becomes a habit.
- In-person meditation classes offer community and support, invaluable assets while healing. Different teachers have different approaches and methods, so don't give up if you don't like the first class. If you find it useful, you might even consider half-day, daylong, or multiday mindfulness retreats.
- There are great mindfulness and MBSR classes online, as well as low-cost and free mindfulness apps. These come with guided audio that take from five to thirty minutes and can be used anytime, anywhere. Find some options on my website, Zoffness.com.

> - I've also listed some great books and workbooks to support your mindfulness practice in the resources section of *The Pain Management Workbook*.

IV. Mood Management

Just as negative emotions amplify pain, joy, relaxation, gratitude, and other positive emotions do the opposite—liberating hormones and neurochemicals that change the body and turn down pain volume in the brain.[7,8] You've witnessed this in this every story you've read so far, and have assuredly experienced it yourself. In this section, I'll teach you three science-backed methods for regulating emotions, including strategies for reducing stress and sadness, boosting pleasure and joy, and inducing affective (emotion) analgesia:

1. Pleasurable activity scheduling,
2. Harnessing the power of sunlight and nature to change the brain, and
3. How to express rather than suppress.

It is very important to note here that "negative" emotions aren't actually bad. People in pain frequently say to me, "I wish I didn't have to feel sad, stressed, or anxious anymore." It's a normal sentiment. But emotions are fundamental to healthy human functioning. Emotions give us information about ourselves, others, and situations. When something scary is happening, for example, the fear we feel flips our nervous system into fight-or-flight mode—saving our lives by encouraging us to run from or fight off the danger. Sadness and grief are signs that something is happening in our lives that requires attention and care. Frustration and anger motivate us to set boundaries and make change.

We don't choose emotions. If emotions were a choice, *nobody* would be sad. Nobody would be anxious or get sick with worry. We wouldn't feel that flash of anger at our child for misbehaving, we wouldn't yell, and we wouldn't feud with our spouse. The reality is that emotions are complex events influenced by our biology, our history, and everything happening in the world around us. It isn't helpful to blame our emotions for existing, disparage them, ignore them, or wish them away. The best we can do for our emotional health is to get to know our emotions, understand what triggers them, and then do our very best to take care of them.

As a first step, take a moment to think about your mood and emotional health in recent months. Overall, have you been feeling happy and joyful? Or has your mood, on average, been low? If you've been feeling sad, depressed, anxious, angry, frustrated, or hopeless more days than not, add "low mood," "anxiety," "hopelessness," and so on to your high-pain recipe. While these emotions are expected when we live with chronic pain, we don't want them to hijack the Pain Dial. Let's now learn some methods of transforming these emotions into low-pain ingredients.

1. Scheduling Pleasurable Activities

Living with pain often means giving up beloved hobbies and activities that bring us joy. This can create resentment, anger, sadness, and grief. Ironically, these hobbies and activities are the very things we need to help us heal.[9] If we peered inside our brains, we'd see that pleasurable activities set off a chain of biological events that release mood-altering neurochemicals known to reduce stress, anxiety, sadness, and depression—while cultivating the very emotions and neurochemicals we need to help lower pain volume. Doing things we love also tickles the parts of the brain and spinal cord that reduce danger messages.[10] Last but not least, pleasurable activities distract our brains, directing our attention away from our bodies and onto other things. Engaging in enjoyable hobbies is thus *especially* important when we live with pain.

ACTIVITY: Cultivating Small Joys: Pleasurable Activity Scheduling

Step 1. In your notebook, make a list of activities that bring you joy. This brief writing activity can help uncover small, hidden sources of pleasure in your day-to-day life. The simplest activities count, as ordinary as buying fresh bread from the corner bakery.

For example:

- Taking a hot shower or warm bath
- Getting a massage
- Dancing in the living room
- Watching a comedy
- Gardening
- Writing a letter to a friend
- Reading a novel
- Having a lunch date with a friend
- Spending time with your granddaughter
- Birdwatching
- Working on the car
- Writing a poem
- Trying a new recipe
- Dining at a new restaurant
- Painting, building, creating
- Playing fetch with the dog
- Attending a concert

Step 2. Cultivate small joys by scheduling one pleasurable activity per day—more if you'd like, but not less—for one week. Specify the activity, the time, and the place you'll do it, and put it in your calendar. For example, *visit the art museum Wednesday @*

12 p.m. Consider inviting someone to join you to increase pleasure and accountability.

Step 3. Each time you engage in a pleasurable activity, check in with your body and emotions before and after by rating your pain level and stress/anxiety level. You can use the traditional 0 to 10 scale (0 = none, 5 = moderate, 10 = severe), or simply write about your experience, noting how you felt physically and emotionally before and after.

PRO TIPS

- The things that bring us pleasure and happiness extend well beyond hobbies and activities. They can be memories, places, thoughts, and moments. For example, an unexpected voicemail, the "word of the day" (Sarah, the wildlife biologist with chronic abdominal pain, listed this as a small joy), memories of your grandfather, or the perfect parking spot. In the same way that something as small and fleeting as a hurried driver cutting you off can ruin your mood, deliberately noting small, good things can lift your spirits and spark joy, too. Make it a practice to start noticing a few good things each day, keeping a list on your phone or in your notebook.
- Another option for cultivating positive emotions is to start your own gratitude practice. This has a ton of research behind it, and is extremely simple to do. Sit down every day at the same time and write three to five things you're grateful for—more if you've got 'em! Start by doing this for just five minutes a day. Make sure to include any improvements in your health or pain as you progress through this protocol. Attending to improvements, no matter how small, offers us some optimism along with the important reminder that pain is ever-changing.

> - If you find this activity challenging, I recommend reading the book *14,000 Things to Be Happy About* by Barbara Ann Kipfer. It has some great suggestions and might inspire you.

2. Harnessing Sunlight and Nature

Sunlight, trees, fresh air, and greenery—these are medicine, too. It's no coincidence that nature sounds, like rain and ocean waves, soothe our mood and put us to sleep. Exposure to sunlight and nature liberates brain chemicals like dopamine, serotonin, and endorphins, chemicals that make us *feel better*. Sunlight also stimulates the production of vitamin D, essential for healthy immune functioning, bone health, and strength. The warmth of the sun, nature's heating pad, helps our muscles relax. Sunlight programs our biological clock, regulates our circadian rhythms, and even improves sleep. Sunlight is so important that people who don't get enough of it can develop a variety of issues, from vitamin deficiencies and suppressed immune functioning to hypertension and depression. It's no coincidence that light therapy boxes mimicking natural sunlight have become a first-line treatment for these conditions. But most of us spend too much time indoors. Thankfully, it's easy to prescribe sun-medicine.

ACTIVITY: Sunlight and Nature

Step 1. Begin by asking yourself the following questions:

- *How many minutes of sunlight do I get each day?*
- *How many minutes do I spend outdoors or in nature?*
- *What small changes to my daily or weekly routine can I make to increase exposure to sunlight and fresh air?*

Step 2. Make any of the following changes:

- Take one ten-minute walk outside every day, bringing along whatever gear you need: a hat for shade, a jacket for warmth, even a walker. If you can't walk, stand outside in the sun. If the sun isn't shining, just get ten minutes of fresh air.
- Eat one meal outside every week.
- Drive to a local park, sit on a bench, and read.
- Ask friends or family to join you for a meal or an activity outdoors.
- When using the strategies in this book, try practicing them outside. Try a body scan outside on a blanket, or diaphragmic breathing while sitting on a park bench.
- When you exercise or pace for pain, do it outside instead of going to a gym.
- Plan one social event outdoors this month.
- Take the dog, your kids, or a friend on a nature hike. Duration doesn't matter; even a short hike will do.

If weather doesn't permit outdoor activity, find other ways to source nature medicine: go for a drive and take in the scenery. Listen to nature sounds, like ocean waves or thunderstorms, using apps like Rain Rain Sleep Sounds or online nature videos. Buy

> indoor plants, put a bird feeder outside your window, lie on the couch and watch the clouds drift by. For those who live in places with dark, cold winter months, sun lamps can be extremely effective at mimicking the biological benefits of sunshine.

3. Express, Don't Suppress

The list of places emotions manifest physically is as numerous as the parts of the body: tension in our shoulders, a clenched jaw, a tight stomach. Take this moment to practice mindfulness, and tune in: Where in your body are you holding stress and tension right now? Our emotions need an outlet, which is why releasing them is known as "venting." Mateo, the child who lost his forearm to a firecracker, discovered this firsthand: Expressing how he was feeling in session finally started relieving his long-suppressed emotional pain.

There are many options for expressing emotions in healthy ways:

- Write or journal
- Yell in the car
- Let yourself cry
- Exercise: Go for a run, lift weights, stretch, do push-ups
- Punch a pillow
- Rip a phone book
- Try boxing
- Create: Draw, paint, dance, perform, sing aloud, play an instrument
- Talk to someone: Call a friend, try therapy (stay tuned for tips on how to find a great therapist!)

Journaling, also known as expressive writing, can be a particularly potent medicine.[11] Start by writing about your pain experience: when it first started, what was going on at that time, treatments you've tried, how they've failed. Dump all of your frustration and angst onto the page. Write about things that have been bothering you, from work to pain to relationships. Vent about your stressors and release your rage. Let yourself feel every feeling. Be as honest as possible without worrying about spelling or grammar. Journaling is a safe and effective way to release pent-up emotions, cultivate mindful awareness, and better know your own mind.

ACTIVITY: Venting

Step 1. Before we can express an emotion, we first need to be able to identify it. Start by answering the following questions:

- *How do you know when you're feeling angry, scared, anxious, or sad?*
- *Where do you feel each of these emotions in your body?*
- *What events, situations, people, or thoughts trigger each of these emotions?*

For people living with pain, one of the most common (and ignored) emotions is grief: grief about the things pain has taken from you, like specific functions or activities, or even grief about the body part that hurts. It's also common to feel anxious: anxious about your body, what's to come, and whether or not you'll heal. In your notebook, list your emotions and what triggers them. For example, "I feel fear when I go to the doctor to see if the cancer has shrunk or not." If this is challenging, try a mindfulness practice. Do a body scan, go inside your body, and investigate. Then try this activity every day for a week: when you feel a negative emotion, grab your notebook and log how intense it is on a 0 to 10 scale, and also document your pain level. Make sure to track all the places you feel these emotions in your body.

Step 2. Once you've labeled emotions and their triggers, you can begin to express them. When you feel a powerful emotion, select a method of venting from the list above and watch your emotions shift and change. You need not vent for long: Aim for fifteen minutes to start. Log how you feel before and afterward, physically and emotionally. If you're new to emotion expression, rotate through the strategies listed above, trying one each day for a week until you find what works best for you. Add your own methods to this list, and make venting a daily practice.

V. Biobehavioral Treatments: Alter Behavior to Alter Biology

Remember the Pain Cycle, that negative feedback loop of thoughts, feelings, and behaviors that keeps our pain system stuck in overdrive? Scientists have created treatments to break these cycles called biobehavioral treatments—"bio" because they alter our biology and neurochemistry, and "behavioral" because they harness the power of behavior change, along with associated thoughts and emotions, to help us do it.

Among the most successful and established of these is cognitive behavioral therapy for chronic pain, or CBT-CP.[ii] This is a specialized, highly tailored form of cognitive behavioral therapy specifically designed to treat chronic pain, improve functioning, and help us return to work, life, and play (similar to how CBT-I is a specific modality for sleep). When done well, it hits all parts of the pain recipe—from our emotional health to our social health to our physical health. Despite common misconceptions,

---

ii The impact of CBT-CP has historically been hard to measure because it lacks a universal operational definition and has no standard protocol. Across studies, CBT-CP is defined, implemented, and measured differently—varying in number of sessions, formats (virtual vs. in-person, individual vs. group), home practices, and techniques used. Even with these inconsistencies, however, CBT-CP consistently shows improvements in both pain and functioning across a wide range of chronic pain conditions.

it isn't a mental health treatment, a treatment for trauma, or a vehicle for processing a painful childhood. Rather, CBT-CP focuses specifically on helping you break your Pain Cycle by targeting the unhelpful behaviors, thoughts, and emotions that keep pain volume dialed high.

CBT-CP has been shown to reduce pain intensity and change pain perception by altering the way our brain processes danger messages from the body.[12,13] One of the ways it does this is via our old friend, neuroplasticity. After treatment, people with chronic pain show changes in the structure and functioning of the parts of their brain that make pain. Because of its success in reducing pain and the need for opioids, doctors have even referred to it as a potential antidote to the opioid epidemic.[14] There's now so much evidence for CBT-CP that it's considered a first-line treatment for chronic pain.[15] It has proven effective for:

- Cancer pain
- Low back pain
- Rheumatoid arthritis and osteoarthritis
- Neuropathic (nerve) pain, including diabetic neuropathy
- Orofacial pain
- Fibromyalgia
- Chronic migraine and tension headaches
- Irritable bowel syndrome
- ...among other chronic conditions.

That we continue to marginalize this treatment as "just psychological" does us all terrible disservice, limiting our options for faster, safer pathways to healing. Other biobehavioral treatments specifically designed for chronic pain include:

- Cognitive Functional Therapy (CFT), a treatment developed by expert physical therapists that targets physical, emotional, and cognitive pain ingredients to address pain and disability. It has had great success for treating pain, particularly chronic back pain.[16] CFT typically proceeds in three stages: understanding

pain, exposure to movement and activity with control, and lifestyle changes.
- Empowered Relief, a two-hour, single-session training developed at Stanford that teaches patients pain management skills. It has been shown to help patients manage pain with less opioid medication.
- Pain Reprocessing Therapy, which is a spin-off of cognitive behavioral therapy. It specifically focuses on cognitive components of pain, correcting catastrophic and negative beliefs.

In the next "Pro Tips" section, I offer some suggestions for finding providers trained in these modalities.

VI. Brain Exercise: Psychotherapy

"Therapy" has become a bad word when it comes to pain. Most people don't want to see a psychologist for pain, in large part because many of us have been told our pain is "all in our heads." Serena Williams's story reminds us that this is true of women and minorities in particular. This dismissal of our experience can make seeking psychotherapy feel risky, as it could seem to inadvertently validate this false narrative.

In addition, we've been taught that pain is purely biological, conditioning us to trust *only* treatments that target physical drivers, and compelling us to steer clear of treatments that address anything other than the part that hurts. But despite the lingering stigma, thanks to modern neuroscience, we know better than our parents and grandparents before us. Given how significantly emotions affect pain, we'd be remiss if we didn't try to harness this power.

There are many different types of psychotherapy to choose from. Of these, cognitive behavioral therapy (CBT) has been found to be the most effective for treating the issues most strongly linked to pain amplification, including stress, depression, anxiety (including obsessive-compulsive disorder, phobias, and social anxiety like Sarah's), and trauma.[17] There are many subtypes of CBT, but in its most basic form, it was developed to improve emotional health. CBT isn't the type of therapy that requires you to lie on a couch and talk endlessly about your mother, although you are certainly welcome to do this if you wish. Instead, CBT offers concrete, actionable strategies and tools for:

- Improving mental health and functioning
- Boosting mood
- Reducing stress and anxiety while offering tools to increase relaxation and calm
- Improving coping and problem-solving skills
- Navigating difficult experiences like chronic pain

The benefits of psychotherapy, regardless of the kind you choose, are plentiful. For starters, it offers us a safe space to vent, process emotions, and increase our EQ (emotional intelligence). It helps us gain insight into our behavior patterns and coping strategies—the ones that work for us, and the ones that don't—along with concrete tools for improving them. Therapy expands our social support network, giving us a trained professional to lean on during tough times. Therapists offer professional, unbiased perspectives and science-backed tools distinct from what a friend can provide.

To correct a common misconception, therapy isn't just for those with mental health diagnoses: It's for *everyone*. Just as exercising our body makes it fitter, stronger, and healthier, so too does psychotherapy, or brain exercise, make our brain fitter and stronger—and, ultimately, better able to manage pain. If it's okay to go to a soccer coach to get better at playing soccer, it's certainly okay to use a pain coach, like a trained psychotherapist, to help you get better at living with pain.

PRO TIPS (FOR SECTIONS V. BIOBEHAVIORAL TREATMENTS AND VI. PSYCHOTHERAPY)

- Try on multiple therapists before settling. Most people hire the first therapist they find. This is a huge mistake, leading to a poor fit, bad experiences, and a subsequent belief that therapy "doesn't help." If we try on ten different shoes before finally buying a pair—and that's just for our feet!—our brains surely deserve a good fit, too. As you meet with different therapists, ask yourself: *Do I like this person? Do I feel comfortable with them, and feel like I can talk to them?* If the answer is no, MOVE ON.
- To find a therapist trained in CBT (for your emotional health—which can, in turn, lower pain volume) or CBT-CP (specifically for chronic pain), ask your insurance company for a list of providers or search for specialists online. There are also links for finding providers on my website, Zoffness.com.
- Therapists come in different forms and levels of training: psychologists (PhD and PsyD), marriage and family therapists (MFT or LMFT), licensed clinical social workers (LCSW), and more. Degrees and years of experience have very little to do with whether a provider is a good fit for you. One of the most influential factors in therapy is the therapeutic alliance, or the relationship: how much you like, trust, and want to talk to your therapist.
- Insurance reimbursement for psychotherapy, particularly as part of a pain protocol, is notoriously difficult to obtain. Insurance companies are infamous for delaying or denying claims to minimize payouts and maximize profits. A written prescription from your medical doctor for CBT and other biobehavioral treatments can help. If your insurance

company rejects your claim, call them and inform them that these treatments have been prescribed by your MD and are "medically necessary." Document everything in writing, including the date and name of the person you're talking to. Then escalate your claim as needed. My patients and I have found that persistence pays off (literally).
- Healthcare Providers
    - Prescribe CBT and other treatments in this book as you would any medication, and write them on a prescription pad. This can help patients obtain insurance reimbursement. It can also help change the culture around pain medicine, increasing the legitimacy of treatments like psychotherapy, biofeedback, and other non-drug interventions.
    - Create a list of local CBT therapists and pain psychologists to whom you can refer, and offer the list to your patients. If you can't find providers locally, look further afield. Any licensed psychotherapist in your state can technically provide virtual treatment to any patient in your state. While teletherapy is less ideal for pain than in-person treatment, it has solid evidence of effectiveness, and is often more accessible and convenient—particularly for those with disabilities.
    - Coordinating care is vital for all healthcare providers: physical therapists, occupational therapists, medical doctors, and psychotherapists. The best way to increase the quality and effectiveness of care, and bridge the gap between physical and emotional pain, is to pick up the phone. While care coordination takes time, these calls are clinically invaluable and usually reimbursable.
    - Pain management workbooks are your friend. They are affordable, accessible, and can be used by nearly any patient with any diagnosis.

13

The Power of Changing Your Mind

COGNITIVE STRATEGIES FOR PAIN

We've seen how cognitive strategies like shifting beliefs and attention can radically change pain. But what if I told you that you've already started using them? When your grandchild skins her knee, you kiss it and distract her with a funny story—stemming the tide of her tears. When you get blood drawn, you make conversation with the phlebotomist to take your mind off that needle. And when your husband heads into surgery, you smile at him reassuringly, telling him that everything will be okay—even if you yourself are terrified. The power of language and distraction to adjust the Pain Dial are legendary, and you've already seen these very real pain-control strategies at work. Now, with the help of science, I'll show you how to harness their power on a much larger scale.

I. The Healing Cycle

Words That Harm, Words That Heal

We've all heard the saying "sticks and stones may break my bones, but words will never hurt me." We know now that isn't actually true.[1] Dr.

Peper, professor and biofeedback expert, showed us how stressful thoughts can tense our muscles, elevate our heart rate, and alter body temperature. We've seen over and over again how nocebic messages, or danger messages, can hijack the brain's emergency alarm, and how hopelessness can trigger stress and suicidality. Mr. A showed us how negative expectations, like his belief that he'd overdosed on a powerful drug (rather than a sugar pill!), can radically change our very biology.

But just as negative, pessimistic, and fearful language has the power to crank pain volume up, hopeful and soothing safety messages have the power to dial pain *down*—changing our brain chemistry and altering our physiology.[2] We saw this in action with Kai, whose hopeful expectation that faux gummies could reduce his pain activated his body's pharmacy, triggering the release of neurochemicals that actually lowered pain volume. We've seen how a combination of beliefs, messages, and other factors can relieve symptoms of very real diseases, from Fabry disease to Parkinson's. It turns out that words that make us feel better *can actually make us feel better.*

These activate what I call the Healing Cycle, which is the opposite of the Pain Cycle—that cycle of chronic pain linking thoughts, emotions, behaviors, and physical sensations you met in Chapter 6. The Healing Cycle demonstrates how these interconnected factors also have the power to work in the opposite direction for our benefit: activating neuroplasticity, stimulating the body's pharmacy, and speeding healing.

One place to jump-start this cycle is by changing the thoughts we think and the messages we believe. For example, this process might begin with replacing an old thought with a different one that might help instead of harm. This language doesn't necessarily need to be optimistic or even positive—even slightly more realistic, science-based messages will do the trick. For example, if Pain Voice tells you, "I'm broken, I'll never get better," triggering feelings of fear, sadness, and hopelessness, we'll replace it with a new thought, like *If the brain can change, pain can change.* This new message inspires new emotions. We might feel hopeful, less stressed, more motivated and reassured.

THE HEALING CYCLE

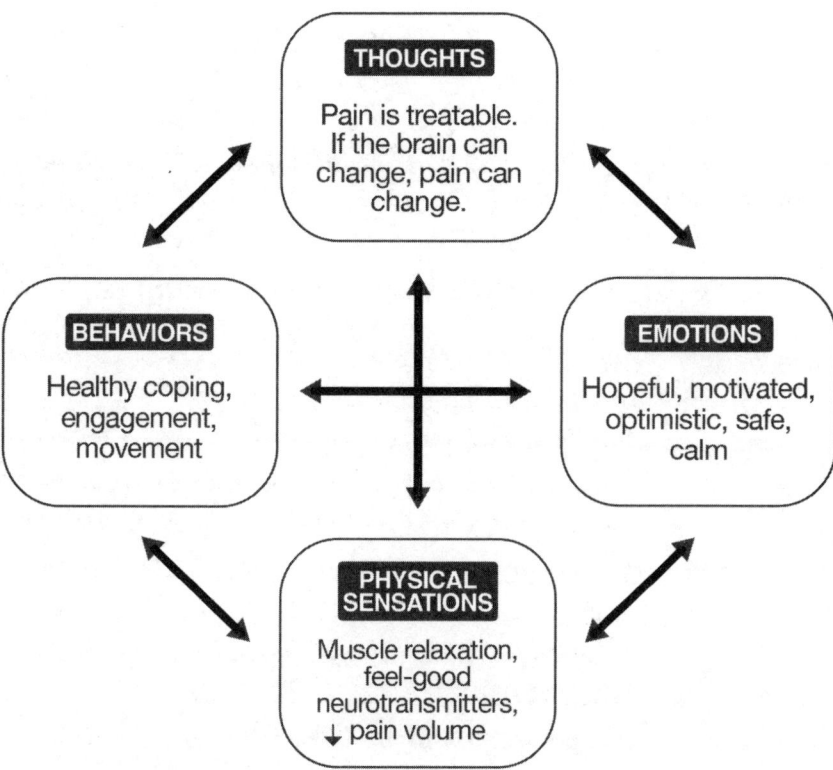

Our thoughts, emotions, physical sensations, and behaviors interact to create a cycle of healing.

Credit: Illustration by Eva Huzella

As a result, our body responds with physical changes. *If the brain can change, pain can change*: the body's pharmacy activates. A flood of mood-boosting, pain-reducing neurotransmitters is released, like endorphins, dopamine, and serotonin. Pain-amplifying stress hormones, like adrenaline and cortisol, start decreasing. *If the brain can change, pain can change*: our muscles relax, energy increases, sleep improves. Our decisions and behaviors subsequently change, too: We might feel empowered to reengage in activities, start moving more, reconnect socially, eat healthier. In this way, *one little thought can completely disrupt the Pain Cycle*.

Here's how the Healing Cycle worked for Fallon, the artist with complex regional pain syndrome:

> Old belief: *I'm broken; I'll never get better.*
> New belief: *Chronic pain is treatable.*
> New Emotions: *Hopeful, reassured, less stressed and anxious, motivated, empowered.*
> New Behaviors: *Engaging in treatment, including physical therapy, occupational therapy, and psychotherapy; scheduling movie nights for social reconnection; making dietary changes; resuming archery, walking, and other activities.*
> New physical changes: *Skin sensitivity lessened, leg lesions healed, circulation improved, strength and mobility increased, sleep and energy improved, pain volume decreased at the brain and spinal cord.*

In this chapter, you'll learn how to activate your body's pharmacy and create your own Healing Cycle using the power of language. I'll teach you how to change the thoughts you think, the words and images you use, and the messages you believe. We'll start by learning how to identify nocebos in your everyday life: words that harm instead of heal.

II. Identifying Harmful Messages: Pain Voice and External Nocebos

Nocebic messages are often hiding in plain sight. But we can't do much to lessen their impact until we learn to identify them. The good news is that once you know how and where to find them, addressing them becomes infinitely easier. These harmful danger messages have two primary origins:

1. *The external world.* External nocebos include any danger messages from the outside world that trigger stress, fear, hopelessness, or pessimism. These commonly and erroneously inform our beliefs about pain—say, for example, that pain always indicates damage, has no cure, or is located exclusively in a body part. Sources include the internet, which confidently and falsely states that chronic pain has no treatment or cure; social media influencers and "expert" websites packed with myths and misinformation; and well-intentioned healthcare providers, friends, and family members who don't understand pain, use catastrophic language, or make negative predictions.

2. *Our internal world.* Internal or self-generated nocebos come in the form of negative thoughts, expectations, predictions, and beliefs. This is your Pain Voice, which you first met in Chapter 6. Pain Voice thoughts are typically negative, pessimistic, discouraging, and even catastrophic. They predict terrible outcomes, amplifying our sense of danger and fear. This is the voice of worries, what-ifs, and worst-case scenarios. Kiran's Pain Voice, for example, incorrectly told her that her pain was due to stomach cancer rather than a kidney stone. Hallie's Pain Voice predicted that pelvic pain meant she'd never be able to have sex, let alone a baby. These thoughts are commonly called "thinking traps," because while they sound true, they're actually false distortions that trap us in a spiral of anxiety, pessimism, and pain. We might even repeat these thoughts aloud, saying them to ourselves and to others, feeding them back to our brain for

processing. This only further amplifies the pain alarm, because *your brain hears everything your mouth says*.

Activate Your Healing Cycle: Catch It, Check It, Change It

My favorite method of dispensing with nocebic messages of all kinds, internal and external alike, is the 3 Cs: Catch It, Check It, Change It.[i]

Step 1, Catch It, teaches us to recognize untrue Pain Voice thoughts and nocebic messages from the outside world as they occur. The second step, Check It, shows us how to test their veracity, like kicking the tires on a car to ensure they're sound. This allows us to answer the question: Is this fact, or false? The third and final step, Change It, helps us transform nocebic thoughts and messages into more helpful, realistic, and hopeful ones.

This, in turn, activates the Healing Cycle: a series of new thoughts, emotions, behaviors, and physiological responses that helps us heal by activating neuroplasticity and our body's pharmacy. People who use these strategies report improved physical functioning, decreased disability, better mood, and *less pain*.[3,4] Once you know how to do it, you can apply the Three Cs in a matter of minutes.

1. Catch It

Much like learning a new word, once you learn to recognize Pain Voice thoughts and nocebic messages, you'll start hearing them everywhere you go. After years of working with patients, I now catch these negative messages on the news, at the hospital, even while waiting in line at the coffee shop. You soon will, too.

The first tell is that, despite initially appearing to be facts, danger messages and Pain Voice thoughts are actually distorted, unhelpful, and untrue. They're often discouraging, predict terrible outcomes, and plant negative expectations. You've seen these with every character you've met in this book so far. Todd, for example, the ultramarathoner who completed 314 miles in just nine days,

---

i The 3 Cs technique originates from cognitive behavioral therapy (CBT).

was told after his car accident that he'd never run again. Joyce was informed that she had the "back of an eighty-year-old," a message that inspired fear and dread. I've listed some additional nocebic messages below, both external (from the outside world) and internal (generated by our brain). Add any that sound familiar to your high-pain recipe on page 180. In the next activity, I'll teach you how to root out and identify your own unique nocebos.

External Danger Messages

"People with this condition have pain for the rest of their lives."

"Chronic pain has no treatment or cure."

"This is only going to get worse."

"It's just bone-on-bone in there."

"Movement is dangerous."

"This is going to hurt."

"You're in pain because your body part (hip, knee, spine) is degenerating."

"Your pain is the result of asymmetry." (Note: Every single body on planet Earth is asymmetrical—including, as you may recall, gold medalist Usain Bolt's.)

Internal Pain Voice Thoughts

"I'm broken, nothing will ever fix me."

"What if I can never [run, play soccer, have sex] again?"

"My back is too fragile, I can't _____."

"Depression and chronic pain run in my family; this is just going to be my life."

"Pain is punishment for _____" (e.g., surviving combat in Iraq when others perished).

"Something is seriously wrong in there, I know it."

"My pain is *only* due to damage or disease."

"I'm a burden to my loved ones."

"This pain will never end."

"My body is failing me."

Unfortunately, it isn't always obvious which messages and thoughts are traps, and which are truth. This next activity will enable you to remove the guesswork. We'll examine both major sources: messages coming from the outside world, and those being generated by your overprotective brain.

ACTIVITY: Catch It: Tracking Danger Messages

Step 1. Answer the following questions in your notebook to generate a list of danger messages you've received in the past. These are important to establish, as they've influenced your beliefs about your body and your recovery. These messages may have come from a book, a healthcare provider, social media, a friend, or a magazine article.

- *What negative, anxiety-provoking, or discouraging messages suggesting danger or harm have you been given in the past? By what sources?*
- *What did these messages sound like? (For example, "Your discs are degenerating," "Your pain will only get worse with time," "There is no cure for fibromyalgia," "This is going to hurt," etc.)*
- *Have you ever felt hopeless about your treatment, prognosis, or diagnosis? What messages or information contributed to this hopelessness?*

Step 2. Next, begin noticing new negative and nocebic messages from the outside world as they happen in real time. Start tuning in, paying careful attention to what you're being told about your pain and potential for recovery. As with the messages above, these may come from a variety of sources. Document them in your notebook as you hear them. Note the who, what, where, and when: *Monday at work, reading an online article in ____ magazine by an expert about how surgery is the only cure for back pain.* Then add these nocebic messages to your high-pain recipe in Chapter 10.

Step 3. Start documenting Pain Voice—your own internal thoughts, negative expectations, and predictions. This is called "thought tracking." To catch Pain Voice, keep your notebook in your bag during the day and next to your bed at night, paying close attention to worried thoughts, negative self-talk, and unhelpful predictions. Pain Voice is often most audible while lying in bed at night. Not coincidentally, this is one of the reasons it can be hard to sleep: When the world quiets down and distractions disappear, Pain Voice often gets very loud. Some questions you can ask yourself to help catch Pain Voice are:

- *What negative predictions about pain or my body have I made? What negative expectations do I have about my future regarding work, relationships, health, or pain? What worst-case scenarios am I imagining?*
- *When I worry, get anxious, or feel sad, what worried or discouraging thoughts do I have?*
- *What what-ifs do I detect? (For example, What if I never get better? What if I can never walk / run / have sex / travel again? What if this pain lasts forever?)*
- *What situations, emotions, people, or events activate negative thoughts and predictions? (Log these carefully; these are the social and environmental contexts that trigger Pain Voice.)*

Step 4. While Pain Voice may speak in your voice, *it isn't actually you*. Once you've tuned in to Pain Voice and know what it sounds like, give it a name, like Petunia the Pain Voice. This technique is called externalizing, a strategy that helps distinguish you from your thoughts, providing some distance and objectivity. Every time you hear a Pain Voice thought in the future, you'll be able to see it for what it is: an overprotective, unhelpful thought generated by Petunia, the well-intentioned but overprotective part of your brain. Add "Petunia the Pain Voice" to your high-pain recipe on page 180 to represent your pain-amplifying thoughts.

PRO TIPS

- If you have trouble tuning in to your thoughts, practice mindfulness to slow the thought process and quiet external noise.
- Journaling and expressive writing, which you learned in Chapter 12 (page 231), is also a great way to capture negative, unhelpful thoughts.
- To get help with thought tracking and the 3 Cs, try cognitive behavioral therapy or Cognitive Functional Therapy. Tips for finding providers can be found on page 237.

2. Check It

Now that we've identified some negative thoughts and danger messages, how do we know for sure they're traps and not the truth? In the next step, you'll investigate the veracity of these thoughts and messages. We'll need to conduct an assessment and gather some evidence. I've developed a series of ten Detective Questions to help you do just this. The answers to these questions will yield rational, factual thoughts—messages you'll use as battle-axes to challenge the pain myths you read by day, and the distorted Pain Voice thoughts you hear at night.

Activity: Detective Questions

This activity helps us get into the habit of *questioning* the thoughts we think and messages we hear rather than automatically believing them. If they're traps rather than truth, the very act of questioning them will poke holes in them, deflating them like balloons. And the answers will guide us to new thoughts that inspire hope and healthy behaviors, and activate our Healing Cycle.

Step 1. At the top of the page, write the thought or message you're investigating. For example, "Your pain will only get worse," as Kai was told about his genetic condition, Fabry disease. It may initially sound true, as this message did to Kai and his family, especially if delivered by a trusted source. And it may actually be true. But what if it's really just a trap? Let's find out.

Step 2. Answer the following Detective Questions in your notebook, writing the answers in full sentences. If the question is irrelevant, skip it. Kai's answers are shown as examples.

1. Q: Is this thought or message a fact? ("Fact" means it is unquestionably, absolutely, and certainly true. If it isn't true, flag it as a likely nocebo.) *Kai's answer: No, it's not a fact that my pain will only get worse. It's a possibility, not a certainty.*
2. Does this thought or message predict bad things? (If yes, it's a nocebo. Predictions aren't facts. Even a doctor's prognosis, which may include expectations about the course of your disease, is a guesstimate—not a fact.) *Kai's answer: Yes, this message predicts bad things. Predicting the future is not a superpower anyone has, even the most skilled practitioners.*
3. What else—neutral or positive—might happen in this situation other than what I'm predicting or being told? *Kai's answer: My pain might not get worse. It might stay the same, it might improve, or it might go away entirely. Even if my pain continues, I might get significantly better at living with it and managing it. If a bad outcome is a possibility, so too is a good one.*
4. How does this thought or message make me feel? (If it makes you feel bad—anxious, stressed, depressed, or defeated—the message is cranking your pain alarm.) *Kai's answer: This message makes me feel powerless, scared, and hopeless.*
5. Is this a safety message, or a danger message? (Danger messages are potential nocebos.) *Kai's answer: This is a danger message.*
6. Is it helpful, or harmful? (If it's harmful, it's likely a nocebo.) *Kai's answer: This message is harmful.*
7. Does it equate pain (hurt) with damage (harm)? For example, "Your back scan shows disc compression / a

herniated disc / degenerative disc disease; this is the cause of your pain," as Joyce was told. (If yes, this is a nocebo. Pain is more than damage alone.) *Kai's answer: Not applicable since this message doesn't reference damage.*

8. Does it contain extreme, black-or-white language? Examples include "all"/"nothing," "everyone"/"no one," "always"/"never," "best"/"worst." (If yes, it's a likely nocebo—the world exists in shades of gray, not black or white.) *Kai's answer: This message predicts that things will* never *get better. This is an extreme, black-or-white message that's unlikely to be true.*

9. What evidence do I have that this might *not* be true? *Kai's answer: I've read cases where people's pain completely changed and improved. Maybe mine could, too.*

10. Is this message likely the truth or a trap? *Kai's answer: This message is a trap.*

Step 3. If you still feel unsure whether this is the truth or a trap after answering the Detective Questions, try entering them into a table like the one below. Kai's example can be found in the first row. I added two of the most common traps below his. On the left, write the potential trap. After completing the ten Detective Questions, enter a summary of the answers on the right, in the Truth column. Seeing the truth in black-and-white is a powerful antidote to distortions and fear. Correcting these beliefs, like the erroneous belief that hurt (pain) is always the same as harm (damage), gives us the power to reinterpret experiences and change the pain we feel.

Trap	Truth
"Your pain will only get worse." (Kai)	It isn't a fact that my pain will get worse. My pain could also improve or go away entirely. If the brain can change, pain can change.
Pain is always a sign of damage.	Bodies can experience pain without damage, and damage without pain. Chronic pain is the brain's danger alarm on high alert.
Pain lives only in the part that hurts.	Pain is constructed by the brain with input from the body and environment. Many factors matter.

3. Change It: Words That Heal

You've learned to Catch It and you've learned to Check It. In this section, you'll learn to Change It: steps for transforming negative thoughts and messages into your *Healing Voice*. This voice gets its power from truth and logic rather than doubt and fear. We'll do this using two cognitive strategies:

1. Cultivating hope and optimism to shift mindset, and
2. Using coping thoughts and safety messages.

Cognitive strategies like these—that change our thoughts, expectations, and beliefs—helped transform Kai's regular cherry gummy bear into a potent pain medication. While pain isn't simply mind over matter, and we can't just think our way out of pain, the following techniques can be incredibly powerful tools for adjusting the Pain Dial. In fact, the new thoughts and messages you'll generate via the following techniques will become important ingredients in your low-pain recipe.

1. A DAILY DOSE OF HOPE

It would be fair to doubt that words matter *this* much. But in a study of more than 220,000 people, an optimistic mindset was significantly associated with better health, less pain, and even a reduced risk of mortality[5]—a finding that has been replicated over and over again.[6,7] Given how much our thoughts impact our bodies, it's important to ensure we're feeding ourselves a steady diet of healthy, hopeful messages. This can be daunting in the face of pain. Here I offer you a shortcut: an antidote to despair, and a powerful prescription you can take every day.

ACTIVITY: The Hope Prescription

Try reciting the following thoughts once a day every day, and more often as needed. These words are important for increasing your sense of agency, power, and control over your body and your health. Make them your new mantras.

- *If the brain can change, pain can change.*
- *I have the power to adjust pain volume.*
- *My body has the capacity to self-heal.*
- *Learning my pain recipe gives me control over the ingredients.*
- *There is a treatment for chronic pain, and I'm on the path to finding what works for me.*
- *I'm moving forward one step at a time.*
- *If pain is always changing, that means that pain can change.*
- *I have hope that things will get better for me.*
- *Every day is a new opportunity to heal.*

> **PRO TIPS**
> - The trick to activating neuroplasticity is to repeatedly rehearse these messages of safety, agency, and hope. To do this, we need to get into the habit of hearing them, reading them, and thinking them. Write them in your notebook, stick them on your fridge, tape them to the wall. The messages we rehearse are the ones we're more likely to believe.
> - Ask your family, friends, and healthcare providers to repeat them, too. When it comes to pain, there's no such thing as having too much hope or self-empowerment.
> - The goal here is not to deny the reality of disease, injury, or illness. Two things can simultaneously be true: You can have a very real, scary disease, *and* you can learn to lower pain volume by helping your brain feel safer and more hopeful.
> - This is also not the same as "magical thinking" nor a suggestion that we can "hope" pain away. However, changing thoughts is a powerful way to activate neurobiological pathways that speed healing, making it a potent form of medicine.

2. COPING THOUGHTS

Coping thoughts are the soothing, reassuring thoughts we use to get through a bad day, especially when Pain Voice gets loud. Coping thoughts help us function and thrive, accomplish goals, and manage flares. These are similar to the hopeful thoughts we generated in the previous activity, with a slight twist: The goals of these thoughts are to increase our ability to cope with pain and cultivate self-compassion. They sound like the encouraging, compassionate words we offer loved ones when they're struggling—and the words that loved ones offer us. For example, think of what you might say to your closest friend when he's struggling, and what he might say to you.

This is a method of delivering ourselves kindness, care, and concern—chicken soup for the soul. These compassionate, loving messages have the power to rewire our brains, adjust neurotransmitters and hormones, and protect against inflammation. There's ample evidence that they can reduce distress, increase our functional ability, and lower pain volume.[8,9]

ACTIVITY: Compassionate Coping Thoughts

Write these in your notebook, then recite them in front of a mirror. To activate neuroplasticity, this strategy requires practice, patience, and consistency.

- *I've had hundreds of pain flares, and I've survived all of them. I'll survive this one, too.*
- *All sensations are temporary. This, too, shall pass.*
- *I am safe. My body is safe.*
- *It's going to be okay. I'm going to be okay.*
- *I'm strong. I can get through this.*
- *I know how to cope with this. Tools I can use to lower the pain alarm are [take a warm bath, distract, call a friend, listen to soothing music, use ice/heat, take medication, etc.].*
- *Living with pain is hard. It's okay to feel sad or discouraged. I give myself grace.*
- *I am a fighter and a survivor.*
- *Just breathe. One day, and one minute, at a time.*

> **PRO TIPS**
> - If you aren't used to feeling hopeful or speaking kindly to yourself, these activities may feel strange at first. But there's wisdom in the adage "fake it 'til you make it." The neurons in our brain are like the muscles in our body: The more we use them, the bigger and stronger they get. Just as the danger-pain pathway gets big and strong with practice and use, so too will the safe-healing pathway grow and strengthen with time.
> - In addition to imagining what close friends or loved ones might say to you in a tough moment, and what you would say to them, you can also source self-compassion from experts via books and guided audio. My favorite self-compassion teachers are Kristin Neff (Self-Compassion.org) and Tara Brach (TaraBrach.com).

III. Guided Imagery

Remember the meditating monks envisioning fire in their bodies as they generated enough heat to dry sheets and melt snow? One of the strategies they employed was imagery, also known as visualization. The human brain thinks in images, making imagery an extremely powerful tool for inducing physiological changes. Try this: Imagine tonight's dinner, whether it be savory pasta with meat sauce, a steaming bowl of soup, or a gooey slice of pizza. Describe the dish in detail, envisioning what it looks, smells, and tastes like. Imagine lifting the food to your mouth. When you start salivating, this is because your brain is using images to anticipate the process of tasting and ingesting—responding as if you're actually about to eat. Indeed, our brains use imagery all the time, even when we're not trying. If you've ever awoken sweaty from a nightmare or had an arousing sexual fantasy, these were the result of brain images altering your body.

But imagery doesn't just affect our salivation, circulation, and erections. It also affects our pain. When Kiran visualized her father's cancer and the way it ravaged his body, her pain alarm instantly amplified. Mental images of mangled, shredded nerve endings in his feet made Kai's pain alarm scream, too.

This same phenomenon occurs in hospitals and doctors' offices around the globe. People with back pain, for example, are often told their pain is due to a "slipped disc." This inspires imagery of a fragile, displaced body part. But what if I told you that the discs in your back are so firmly attached to your vertebrae that they can never, ever slip? That's the truth. Yet these commonly used images feed the brain inaccurate, inflammatory data that ratchets up the pain alarm.[10,11] Real images can amplify pain, too, which is why simply viewing an "abnormal" MRI of your back can actually make your pain worse—even if it only shows standard, age-related changes.[12,13]

This phenomenon occurs because the same parts of our brain are activated when envisioning or imagining pain as when actually experiencing it.[14,15] The rest of our body responds, too, from our muscles to our hormones to our immune response.[16]

However, we can also harness the power of imagery to dial pain down. Kiran experienced this firsthand, her pain diminishing the instant she saw the image of the kidney stone on the screen. Guided imagery, or visualization, is a method of inducing physiological changes in the human body using mental imagery. Imagery has proven effectiveness for a variety of health conditions: alleviating nausea, reducing blood pressure, easing the tremors of Parkinson's disease—and reducing all kinds of pain, including cancer pain.[17] Now, we'll use it to help adjust *your* Pain Dial.

ACTIVITY: Self-Healing Imagery

If imagining discs slipping and cancer growing can make the pain alarm scream, what would envisioning the opposite do? Here you'll learn to apply anti-inflammatory imagery just as you do creams and salves. Self-healing imagery is a practice that uses visualization techniques to envision your body healing from illness, injury, and pain. It combines relaxation, breathwork, and cognitive strategies to support the body's natural healing processes. I regularly teach it to my patients with great success. One of my patients, Tom, a firefighter, used this technique so successfully to end his chronic abdominal pain that he brought it back to the firehouse and taught it to his entire crew.

Step 1. To begin, pay close attention to the images you have of your pain, and the words you use to describe it. Can you "see" the pain in your body? What have you been told about its source? What do you imagine is going on in there? I always listen closely for the pain image: the slipped disc or compressed spine, the "bone on bone," the cancer instead of the kidney stone. What dangerous images are conveyed by the pain story you carry? What thoughts, beliefs, and emotions does it inspire? If the image makes you feel afraid, hopeless, or debilitated, it's one we urgently need to change.

Step 2. Lie down in a quiet place where you won't be interrupted. Turn off your screens, put them away, and close your eyes. Check in with your body and emotions, noting your pain rating (0 = none, 5 = moderate, 10 = severe) and stress or anxiety level (0 = none, 5 = moderate, 10 = extreme). Practice diaphragmatic breathing until your body starts to relax.

Step 3. Travel inside your body, focusing on the place that hurts. Examine your pain, describing it in detail: color, temperature, size, shape, and texture. Notice if it's dark or bright, hot or cool, heavy or light, static or moving. This is your *pain image*. When you peer inside, this image may be the same as in step 1, or it may be different. This is normal.

Step 4. Now imagine this part of your body pain-free. When healed, what color, temperature, size, shape, and texture do you envision this body part would be? If it was red-hot before, is it now blue and cool? If it was sharp and throbbing, is it now smooth and still? This is your *healed image*.

Step 5. Activate your imagination. What healing process is needed to transform your pain image into the healed image? This transformational process can involve a change in color, temperature, size, speed, or texture. You can transform the pain from hot to cold, send an icy waterfall down your leg, change its color from orange to blue, shrink it or grow it, offer it soothing sunlight or a deep massage, change its shape and texture. Whatever you need to do to change your pain, imagine you can do it.

Step 6. With eyes still closed, send this transformational healing process to the body part that hurts, picturing your pain changing. Imagine that this self-healing process is actually working, and that you are transforming your pain.

Step 7. Slowly bring your attention back to the room. Check in with your body and emotions, noting the ways in which sensations and emotions changed. Each time you use this practice, look inside your body anew to see what it needs.

PRO TIPS

- To practice imagery, it's best to first use a guided audio recording in which someone guides you through the steps—rather than practicing on your own. You can find some recommended options for guided audio on my website, Zoffness.com.

> - For additional support and guidance, find a healthcare provider trained in visualization or guided imagery. This can include psychologists, social workers, nurses, body workers, and other healthcare professionals.

IV. Distraction Strategies: Diverting the Gorilla

A few days ago, I removed a hot pan from the toaster oven using an oven mitt. When my phone rang, I removed the glove to answer it, leaving the pan on the counter. Distracted by the conversation, I reached for the hot pan and grabbed it with my bare fingers. It burned like hell. I quickly dropped the pan. My fingertips were red, the skin already beginning to welt and pucker. (If you're wincing, those are your mirror neurons helping you "feel" my pain.) I ran them under cold water, then applied petroleum jelly and bandages. But the pain was relentless. My fingers burned and throbbed. I'd been invited to a friend's birthday party that evening and didn't want to cancel, so I grabbed my coat and ran out the door. During the cab ride, the pain was all-consuming. But as soon as I sat down at the restaurant with friends, immersing myself in conversation, food, and laughter, the pain disappeared. Not just a little bit, but completely. It was only when I got back in the taxi to return home that my fingers started throbbing again.

As we saw in Chapter 6, distraction pulls the brain's magnifier, the prefrontal cortex, off the body part that hurts, significantly reducing pain perception. This doesn't help only when pain is acute and passing, like mine; it can also reduce the intensity and severity of chronic pain.[18] As Kamal experienced in his yard while absorbed in happy memories and playing with his pup, distraction strategies have evidence for reducing the pain of chronic conditions from rheumatoid arthritis to cancer.[19] There's a good chance they can help you navigate pain flares, too.

Distraction strategies come in many forms. The one thing they share in common is that they have the power to draw our brains away from our gorillas, making them a bit more invisible. They can be *activities*, like hobbies or movement; they can be *cognitive*, strategies that engage your brain in a difficult task or use your imagination; they can be *sensory*, involving touch or sensation (more on this in the next chapter); or they can be *emotional*, engaging your emotional brain to change your pain. The best kinds of distractions, like the one I experienced at the restaurant with friends, are multisensory, engaging multiple parts of your brain at once. Below I'll guide you through the steps for creating your own personalized Distraction Plan: a protocol you'll have at the ready in case of a flare, much like the pain medications you keep stocked in the medicine cabinet just in case.

ACTIVITY: Crafting a Distraction Plan

Step 1. In your notebook, answer the following questions:

- *When have you focused on pain and noticed it started feeling worse? For example, lying in bed at night without other distractions, when you're home alone, or when repeatedly asked about it?*
- *What about the opposite? List a few times you were so absorbed in enjoyable or engaging activities that pain seemed to fade into the background. For example, while laughing at a funny movie, playing with your kids, or engrossed in a favorite hobby.*

Step 2. The following events tend to focus attention on pain and amplify pain volume. As you read through these, consider how you might start minimizing them:

- Focusing on pain by talking about it, thinking about it, or worrying about it

- Having people constantly ask about pain and symptoms
- Staying home and missing work/school/events because of pain
- Tracking symptoms and pain rating for months on end (depending upon your diagnosis, a few weeks to a month should be more than enough time to establish a basic pattern)

If you notice that these ingredients might be contributing to your pain, add "focusing on / thinking about / talking about / tracking pain" to your high-pain recipe.

Step 3. Consider which of the distraction strategies below might work for you.

- Distract with *activities and hobbies* you enjoy. These could be hiking, tinkering in the garage, baking bread, spending time with friends, discovering new recipes, watching movies, playing basketball. Write these in your notebook.
- Distract with *cognitive tasks*. What mental exercises absorb your brain and take your mind off pain? Perhaps you enjoy reading books, writing, creating art. If you like games and brain teasers, consider Sudoku or crossword puzzles. Build a model airplane, listen to an audiobook, research your family tree. Take an online class and learn the constellations or how to fix your watch. Try guided imagery. Make a list of ideas.
- Distract using your *five senses*: sight, sound, taste, touch, and smell. Rub or touch the spot that hurts, and notice how that changes pain. Get a massage. Apply an ice pack or heating pad. Take a hot bath or cold plunge to stimulate touch receptors in your skin and engage your brain's somatosensory cortex. Listen to music or make your own.

> Go outside and absorb the sights, sounds, and smells of nature—a known muscle relaxant and pain reliever. Add your favorites to your list.
> - Distract using *emotions*. Different emotions can generate different sensations. Consider things that change your mood, generate feelings of joy, or make you laugh. It could be a favorite comedian, a funny friend, comic strips, or movies. What emotions captivate your attention? Try suspense: Engage your brain by listening to a suspenseful podcast or reading a murder mystery. Finally, list activities that help you soothe or relax. It might be calming music, yoga, diaphragmatic breathing, or mindfulness.
>
> **Step 4.** Craft your Distraction Plan by selecting a few distractions from each category, considering which might work for you in a pinch. If you have an upcoming medical procedure, for example, plan to bring music and headphones and line up distracting videos. If evenings are particularly tough, plan to end your day with a warm bath while listening to a podcast to help distract your brain and ease pain.

Tips for Healthcare Providers: *You* Are the Placebo

Given the importance of words, this is an especially valuable topic for healthcare providers. Here are some tips for using cognitive strategies and language to better help your patients. In conjunction with the other tools offered in Part III, they can significantly improve treatment outcomes. As healthcare providers, we—how we present information, the words we choose, even our tone of voice—are a critical part of the medicine we prescribe.

- As much as possible, avoid catastrophic, scary, or inflammatory language. Fear is a notorious pain amplifier.
- Avoid predicting the future beyond offering diagnoses and necessary prognoses. In particular, minimize negative predictions. Make sure to disclose that you're working with percentages and possibilities, not absolutes. Always leave room for hope.
- Never tell patients that their pain is incurable. While a *disease* may be incurable, a "disease process" and the "pain process" are two distinct things. An incurable disease doesn't necessarily mean incurable pain, as Kai's Fabry disease demonstrates.
- When it comes to treatments and medications, emphasize the rarity and unlikelihood of side effects and adverse outcomes if they are indeed rare. Even with drugs as common as statins for cholesterol, research shows people are more likely to experience potential side effects like muscle pain if you suggest they should expect it.[20]
- Offer reassuring words. Inform your patients that research supports pain's treatability, and offer basic pain science in just a few sentences. For example: "Pain is constructed by the brain, which is constantly changing. This change process is called neuroplasticity. Science shows that learning to harness the power of neuroplasticity can help us rewire the pain system. Here are some tools and resources to get started."
- It's better to describe procedural information, the purpose of the procedure, or the general sensation someone might feel than predict they'll experience terrible pain. As we've all seen, a 10/10 on the pain scale for one person is a 4/10 for another. Moreover, planting expectations that trigger anticipation anxiety is a notorious pain amplifier. Consider:
    - "First, the nurse will clean your arm, then you'll feel the cold alcohol pad."
    - "I'm going to inject an anesthetic to numb the area so you'll be comfortable during the procedure."

- "This may feel like a warm sensation."
- "People usually feel some pressure."

- Some procedures hurt—there's no way around it. But there are many strategies you can employ starting now to help reduce acute procedural pain and discomfort during office visits. The best of these use a combination of cognitive, emotional, and social tools to distract and soothe the pain alarm. They begin with the language you choose, which can be extremely simple:

  - "It will help if you rub or massage your skin in this spot."
  - "To increase your comfort, take deep belly breaths, relax your muscles, and turn your arm into a limp noodle."
  - "I'm going to ask you some questions to distract your brain."
  - "With your permission, I'll touch your arm as a soothing mechanism." (I've seen ob/gyns successfully employ language- and touch-based strategies before pelvic exams, and even a dentist who hired a massage therapist to provide foot massages during painful dental procedures.)
  - "Here's a ball to squeeze."
  - "When you come in for the procedure next week, make sure to bring headphones, music, and podcasts to help you distract and relax." (I always recommend that my patients bring headphones, music, and other distracting, self-soothing tools to every anxiety-provoking procedure and appointment. It's incredibly effective.)

- Verbal suggestions also help lower the pain alarm, like these:

  - "Imagine a soothing, cool sensation washing over the area where you feel discomfort, like waves on a beach."
  - "With each exhale, allow the discomfort to gently release and fade."
  - "Take three big breaths, and with each exhale, imagine blowing the feeling away from your body."

This is known as "hypnotic analgesia," and it works.[21,22] Around the globe, there is an entire corps of healthcare providers trained in medical, or clinical, hypnosis, including the physician-founded American Academy of Clinical Hypnosis. This technique desperately needs rebranding, as people rarely want to be "hypnotized." But this is a misnomer. Simply put, clinical hypnosis is a method of employing language to alter physical and emotional experiences like pain. It's an invaluable tool, one worth learning if you have the time and resources.

- Empathy is an elixir. In a study of approximately 1,500 patients with chronic pain published in the *Journal of the American Medical Association*, those treated by highly empathic providers reported significantly better clinical outcomes—less pain, better functioning, improved quality of life—compared to those treated by less empathic doctors.[23] Studies have even shown that empathy and warmth can make medications, like creams for rashes, more effective. Empathic providers validate emotions, ask questions, listen attentively, and express concern. Other tips include making eye contact and calling patients by their name. A few effective ways of expressing empathy verbally include:

    - "Your pain is real."
    - "I believe you."
    - "It sounds like this has been a long and frustrating journey."
    - "I can understand why you're scared/worried/angry."
    - "I hear that you're suffering."

- When working with people in pain, ask yourself (and your patient):

    - *What does this person need in order to feel hopeful?*
    - *How can I inoculate them against hopelessness while still being honest and accurate?*
    - *How can I increase my patient's buy-in and motivation to pursue effective, biopsychosocial treatment and care?*

- Refer, refer, refer. A multidisciplinary treatment team is your best friend. Have a list of physical therapists, occupational therapists, biofeedback providers, manual therapists, trauma treatments, CBT providers, and other psychotherapists handy to whom you can refer. For children and adolescents, child life specialists are invaluable, underutilized assets.
- *Prescribe this book.*

14

Social Medicine

BLOOM WHERE YOU'RE PLANTED

If social factors determine our health, isolation shortens our lifespan, and disconnection hurts, how can we effect change? So much seems out of our control: We don't choose the environment in which we grow, our parents' socioeconomic status, or the safety of our childhood homes. We don't choose the color of our skin, our sex, or how we're treated because of it. We don't choose our family, nor the genetics, abuse, or dysfunction that come with it. And sometimes, loneliness is simply a fact of life. We're told to "bloom where we're planted," but the seed doesn't always get to choose its soil.

That said, there are many social and environmental factors we *can* control—especially as we grow. We can choose the people we surround ourselves with, selecting friends and romantic partners who are loving, supportive, and kind. We can lean heavily on our communities and ask for help, as Coach Murph did during his cancer treatment. We can identify motivational role models and peers, as Mateo did when navigating phantom hand pain. We can set boundaries around unhealthy relationships. We can limit our exposure to environmental stressors, like the news and social media, and reduce our consumption of pseudoscience, choosing to instead consume high-quality resources. We can be aware of implicit biases, become our own advocates, and clamor for better, more equitable healthcare. And we can break cycles of abuse by treating our *own* trauma, as Hallie did when she moved out of her parents' violent home and sought help. In this chapter, you'll find a protocol for implementing your own social medicine for pain.

I. We Heal in Community

In all of the stories I've shared with you so far, social connectedness played an important role in recovery. This is no coincidence. Social support, safe touch, and healthy relationships are scientifically proven to reduce pain—so much so that some healthcare providers, myself included, believe we should be prescribing social medicine as frequently as we prescribe pain medications.[1,2] This is particularly important for people living with chronic pain, who often experience reduced or restricted social lives.

Social interaction may feel particularly challenging or intimidating if your social muscle has atrophied from disuse, or if social events trigger pain or anxiety, as they did for Sarah, the wildlife biologist with chronic abdominal pain. But this is eminently treatable. The next exercise will help you establish comfortable goals and give you tools to progress with confidence. There are countless ways to increase your daily intake of social medicine. In the following brainstorming activity, we'll identify options that will work for you. The most important thing to remember is this: Pain doesn't get to dictate your social life or your happiness.

You do.

ACTIVITY: Social Medicine

Step 1. If you've stopped or restricted social activities because of pain, or have been feeling lonely, isolated, or unsupported, add this ingredient to your high-pain recipe on page 180. Continue your investigation by answering the following questions:

- *Do I feel lonely or isolated? Could I benefit from additional support?*
- *Do I have a reliable, robust social support network? If so, who's in it (e.g., family, friends, religious groups, clubs, etc.)? If not, continue answering the questions below.*
- *How often do I attend or organize social events on a weekly basis?*
- *What social activities have I stopped or reduced since pain began? Which of these do I miss and want to get back to?*

Step 2. Brainstorm ways to increase healthy social interactions that suit your levels of comfort and functionality. Murph and Rose engaged their entire community during Murph's cancer recovery, mobilizing them to visit, send love, and bring food. Mateo, who'd been socially isolated for months after losing his arm, bolstered his social support via trustworthy healthcare providers and role models like Emmett. Hallie joined an outdoor yoga class. As part of your brainstorm, consider the following questions:

- *How can you increase social support while reducing isolation? Which friends, family members, or community members would you enjoy spending more time with?*
- *Bolster social support via activities you love. What do you enjoy doing? Make a list: jazz piano, building, birding, programming, pub trivia, pickleball.*

- *How might you protect time to engage in more of these activities with others? (For example, join a book club, chorus, or hiking group; register for a pottery or naturalist class; join an online support group; schedule weekly Zoom dates, join a gym.)*
- *How can you leverage your networks, community resources, and the internet to generate ideas? What opportunities for social interaction appeal to you? Do some research, and make a list of community-based resources that offer easy access: the corner café, library, animal shelter, church, neighborhood associations, local speaker series, soup kitchens, senior centers, YMCA, or JCC. Look up local classes, online courses, volunteer activities, and community gatherings. Consider trying therapy for professional-level social support. If you have children, can you take them to the playground to interact with other parents, or join a school committee? Can you block off time to take the dog to the park and talk to other dog owners?*

Step 3. Review your list of social medicine options, ranking them from 1 (easiest) to 5 (hardest). Circle the 1s and 2s; these are the ones we'll start with. If you're an extrovert who typically finds social engagements easy, begin wherever you like. Select one or two activities to try this week, noting where, when, and how you'll accomplish them. The smaller, more realistic, and more specific your goals, the more likely you are to achieve them. For example, *Tuesday at noon, invite Cam to lunch. Saturday at 9 a.m., fill out volunteer application for the American Red Cross. Thursday at 4 p.m., drive the kids to basketball practice and talk to one parent. Sunday at 8 p.m., send email invite for neighborhood happy hour.* This is the foundation of your social medicine plan.

> **Step 4.** Once you've successfully achieved a few 1s and 2s, try a few 3s and 4s. The more you try, the easier they'll become. Gradually build from one social activity per week to one per day, even if it's just saying hello to the cashier at the grocery store.
>
> **Step 5.** The only person who can prioritize your Social Medicine plan is you. Each week, outline a few social activities you'd like to try in your notebook, along with the who, what, where, when, and how.

Fuzz Therapy

Companionship can also come in four-legged forms. If you're a pet owner, you already know the comfort and joy animals can bring. This "fuzz therapy" provides not just emotional benefits, but physical benefits, too. Having a pet can:

- Bump mood-boosting neurotransmitters
- Ameliorate anxiety, depression, and loneliness while increasing pleasure and joy
- Reduce immune-suppressing stress hormones
- Lower blood pressure
- Serve as an anti-inflammatory agent
- Boost brain levels of natural opioids
- Reduce pain severity and intensity[3,4]
- Increase our ability to cope with pain
- Significantly reduce our need for pain medications.[5]

This is one of the reasons therapy dogs have become increasingly welcome guests in hospitals and other healthcare settings. Petting our pets, particularly the fuzzy kind, also stimulates the release of oxytocin, an important brain chemical that facilitates social bonding and feelings of connectedness. Animals that need to be walked can be a particular boon for people living with chronic pain, getting us out of the house and into the sun, motivating us to move, and introducing us to neighbors.

SOCIAL MEDICINE 273

Credit: Calvin & Hobbes © 1986 Watterson. Reprinted with permission of Andrews McMeel Syndication. All rights reserved.

PRO TIPS

- Sometimes, we don't reach out when we need support because we feel ashamed, weak, burdensome, or worry we'll be judged. But everyone needs social support—it's simply how we're designed. And we need it even more when we're in pain. Despite our negative predictions (remember, that's just Pain Voice!), most people actually *enjoy* offering support. Coach Murph learned this firsthand as he watched his entire community show up for him and his family. When we get to show up for others, it gives us a sense of purpose and fosters feelings of connectedness. It's also biologically rewarding, releasing feel-good brain chemicals like serotonin and oxytocin, and decreasing stress hormones. Ultimately, friends, family, and healthcare providers don't know we need help or connection unless we tell them. Reaching out isn't only acceptable, it's critical.
- Digital interaction isn't a proxy for human interaction, nor the health benefits it offers. Some forms of online social connection, like social media, can actually *increase* feelings of depression and isolation. In fact, the more time we spend scrolling social media, the lonelier we report feeling. There are other, better digital resources out there.
- For example, Meetup.com is a website dedicated to creating opportunities for in-person and virtual gatherings, bringing together people with similar interests, from bowling to beading.
- For those who are immobilized or can't leave the house, consider identifying some online communities that bring you pleasure and joy. Take an online art class, audit a

university course, join a virtual book club or networking group.
- If you want to join a pain support group, select online communities that offer science-backed strategies and don't engage in fearmongering (whether intentionally or inadvertently). Reject groups and chat rooms spewing misinformation, panic, and horror stories—they don't help and only spike your pain alarm.
- To get some fuzz therapy, consider adopting a pet. Animal shelters are overflowing with cats and dogs in need of safe homes. If you can't have a pet long term, you can temporarily foster puppies, kittens, and adult animals. You'll help them socialize, then give them back after a few weeks or months. Shelters also always need volunteers to visit and give their animals some company, cuddles, walks, and love. You'll get some in return.
- There's a strong correlation between social anxiety and chronic pain.[6] Having read this book, I suspect this doesn't surprise you. Anxiety is a notorious pain amplifier, as are the loneliness and isolation that avoiding social situations can create. If, like Sarah, you tend to experience flares before social gatherings, parties, dates, work, or school, social anxiety is a likely ingredient in your high-pain recipe. A CBT therapist can help. Find tips for finding one in Chapter 12.
- Loneliness is particularly common among those who have recently experienced a loss or split, have a toxic family system, or are estranged from family members. This can be exacerbated during the holiday season between Christmas and New Year's, a time when people are typically with loved ones. One antidote is to deliberately plan social

events or travel over the holidays to distract, soothe, and get you over the hump. This can be a creative project with friends, a trip abroad, a local camping trip, even a dinner party.
- Family estrangement—either via the silent treatment, ostracization, or another form of severed communication—is rarely talked about due to shame and stigma. However, it's more common, and more harmful, than you might think. One in four of us is reportedly estranged from our father, and that number increases significantly when we include estrangement from mothers, siblings, and other family members.[7] Known as a "social death," estrangement is a potent high-pain ingredient. For example, Joyce's estrangement from her son was a hidden ingredient in her pain recipe, serving as a major source of grief and stress. If it applies to you, add it to your recipe on page 180 if you haven't already. As an antidote, double down on boosting social support via the activities presented here. There are also entire communities, support groups, and books dedicated to people experiencing estrangement.[i]
- Organizations dedicated to facilitating social connection include the Coalition to End Social Isolation and Loneliness (EndSocialIsolation.org), the Foundation for Social Connection (Social-Connection.org), Commit to Connect (CommitToConnect.org), and Project Connect (ProjectConnect-US.com).

---

[i] My favorite book on the science of estrangement is *Ostracism: The Power of Silence* by Kipling Williams, who has also published dozens of scientific articles on this topic.

II. Touch Medicine

Touch can be powerful medicine—one we don't use nearly enough. While the safe touch of others can be soothing, we can also activate these mechanisms ourselves. In fact, if you've ever shaken your hand after smashing it or rubbed your head after bashing it, you already have. Rubbing, light scratching, gentle stroking, massage, movement, and vibration all disrupt the transmission of pain messages between brain and body, changing the pain experience. Touch also triggers a cascade of biological changes inside our bodies. It:

- quiets the fight-or-flight stress response.
- reduces muscle tension.
- lowers blood pressure and heart rate.
- increases feelings of safety.
- boosts immune function.
- releases endogenous opioids and oxytocin, brain chemicals that play a fundamental role in social bonding, comfort, and pain relief.
- inhibits danger messages between body and brain.
- soothes our threat system to reduce pain.

Scientists have hacked this knowledge to develop some (vastly underutilized) pain management devices, like the transcutaneous electrical nerve stimulation, or TENS, machine: a small, battery-operated device that delivers adjustable, vibrating electrical impulses to the skin. TENS units are generally safe and noninvasive, have no risk of overdose, and have evidence of effectiveness for a long list of chronic pain conditions, from fibromyalgia and spinal cord injury to neuropathy and knee pain.[8,9] We can safely use them on ourselves, by ourselves.

Devices for children that utilize touch medicine, like Buzzy the Bee, are renowned for reducing and even eliminating needle pain during injections.[10] Buzzy Pro, the adult version, has been successfully used for dialysis, blood draws, and long venipuncture procedures. A similar, less-expensive alternative is called ShotBlocker.[11] To add to our list of options, temperature changes on the skin can also change pain: Heat and cold modify sensations while disrupting the flow of danger messages between body and brain. Sarah, lover of biology and big words, discovered this firsthand when she used touch-related tools like heating pads and hot-water bottles to soothe her abdominal pain. All of these touch-related strategies for pain management are relatively inexpensive and exceedingly uncontroversial, and there's no reason every primary care doctor, pediatrician, anesthesiologist, and pain specialist shouldn't use them.

Touch science also helps explain why certain movement-and-touch therapies like massage therapy, physical therapy, and occupational therapy can be helpful for pain.

Depending upon the type of pain you're experiencing and your unique brain, the following strategies can be helpful:

- Heating pads and ice packs
- Warm baths and showers
- Cold plunges and ice baths
- Hydrotherapy (water immersion)
- Massage
- Applying pressure or soft stroking
- Products like Icy Hot, Aspercreme, Bengay, arnica gels and creams, and Tiger Balm, which use a variety of chemicals to create numbing, cooling, and/or warming sensations (with a caveat to track side effects and other changes)
- TENS units
- Buzzy the Bee, Buzzy Pro, ShotBlocker

PRO TIPS

- If you're partnered, married, or have sources of safe touch, ask for touch medicine. Request a belly rub or a gentle shoulder massage when you're in pain, and see if it helps. Spend more time holding hands, cuddling, and reading to your kids while they're in your lap. Give and receive massages and back scratches. Insignificant as these activities may sound, they stimulate the touch fibers in your skin and activate the brain chemicals that lower the pain alarm and help us *feel better*.
- People who are isolated, unpartnered, or don't have close families may have insufficient touch in their lives, or experience touch deprivation. There are many ways to increase touch medicine outside of families and romantic relationships. Massage therapy is one option. If insurance will reimburse it or you can otherwise afford it, it may be worthwhile to book regular, monthly massages. Other options include Somatic Experiencing therapy, a body-centered approach to treating pain and trauma that you'll learn more about later in this chapter; animal-assisted therapy, which facilitates therapeutic interactions between humans and service animals like dogs and horses; and even professional "cuddle therapy," which involves consensual, carefully curated types of safe touch with trained professionals (e.g., CuddleSanctuary.com—it's a thing).
- Using heat and cold, as Kamal did for his arthritis and Sarah did for her stomachaches, are quick, free, and easy methods of changing sensations on your own. Try heating pads, cold compresses, warm baths, and cold plunges to see which work best for you.

- Of note, not all kinds of social touch feel good. As with all sensations, context and emotions matter, too. While safe, pleasant, affiliative touch—like holding hands with a loved one, a reassuring squeeze from a friend, or a pleasurable back scratch from a trusted scratcher—will lower the pain alarm, touch that is aggressive, unwanted, or threatening will instead *amplify* danger messages. Trust your gut. If touch feels unsafe, uncomfortable, or icky, say no and end it.
- <u>Healthcare providers</u>: Consider using and recommending TENS units, ShotBlockers, Buzzy Pro, and/or Buzzy the Bee. These are particularly helpful for acute pain and procedures, and TENS units can also provide relief from chronic pain. If of interest, you can also learn how to implement forms of consensual, safe touch before and during procedures, like the ob/gyn in the previous chapter who, with permission, offers safe touch during procedures to distract and soothe, and the dentist who offers foot massages in his clinic during cleanings and procedures. Your clinic or institution might consider the same. Not only will it benefit your patients, but it can also increase your efficiency—saving time and money.[12]

III. Better Boundaries: Restoring Safety

Relationships are critical, but they aren't always healthy. If any of your relationships are abusive verbally, emotionally, physically, and/or sexually; your loved ones hurt you; or yours is a toxic family system, these are ultimately bad for your brain—and your pain. Unhealthy interpersonal dynamics can be codependent, manipulative, or controlling, or characterized by dishonesty, conflict, and gaslighting.

One of the most powerful tools we can employ in our relationships, both healthy and unhealthy alike, is boundary setting. Boundaries erect safe guideposts and parameters around unhealthy relationships, and make healthy relationships healthier. By protecting your energy and resources and helping you feel safe, they can lower the pain alarm. This skill, while rarely taught, can be life-changing when used.

> ## ACTIVITY: How to Set Boundaries
>
> Try this four-step activity as a starting point for increasing your sense of safety and establishing healthier boundaries. In your notebook, reflect on how each applies to you and your social health.
>
> **Step 1.** *Identify your needs.* You can't set boundaries until you know what your relationship needs are. What kinds of words, behaviors, and interactions with others help you feel safe, calm, and secure? Consider physical, emotional, sexual, and even financial boundaries. How do you expect others to treat you? What does being spoken to kindly and respectfully sound like? Conversely, what words and behaviors do you consider abusive, toxic, or boundary violations?
>
> **Step 2.** *Assess current boundaries.* Are the boundaries in your current relationships healthy? Consider who you spend time with, how they treat you, what you need from them, and whether you're getting it. Who are your safe, supportive relationships, and what makes them safe? Which relationships are less safe, unsafe, or toxic, and what interactions make them so? Consider the people who give you energy, and those who are draining. How can you increase time with gains and decrease time with drains?
>
> Distinguishing the people with whom you have healthy boundaries from those you don't can be challenging: Sometimes the

people we love the most are the people who violate our boundaries the most. Make a list of people who tend to violate your boundaries and the ways in which they do so. It might be coworkers, friends, your parents, or your children. Who continues to make demands even after you say no, is manipulative, or administers the silent treatment when they want something? Add abusive, unsafe, or unhealthy relationships to your high-pain recipe on page 180, and safe, nurturing relationships to your low-pain recipe.

Step 3. *Use your words: Learn to say "no."* An important part of setting boundaries is learning to use your words assertively and confidently. This often involves learning to say "no." Nos aren't as black or white as they might initially seem; rather, they exist on a scale from "hard" to "soft." You get to choose which kind you'd like to use, and when. Hard nos are definitive and finite, and sound like this: *Sorry, I already have plans. No, I can't make that deadline. Please stop talking to me that way.* Hard nos can also clearly communicate the consequences of violating your boundaries, as we (hopefully) do with our children: *Please put down your phone. If you haven't done so by the count of three, I will need to take it away until tomorrow.* On the other end of the spectrum are soft nos. These are gentler, and even leave room for a potential yes in the future: *Maybe later. I have to check my calendar. I'm tired right now, but ask me in an hour.* Make sure you know what kind of boundary you want to set before you set it. Then offer it clearly and kindly.

Step 4. *Change how you spend your time, and with whom.* Just as you can set boundaries with words, you can also set boundaries with behaviors. Help your brain and body feel safer by spending more time with people who are kind, calming, supportive, and encouraging, while decreasing the amount of time you spend with people who are abusive, draining, or cruel. Your time is yours, and it's valuable. Which relationships need better limits, and how can you set them? Do you need to tell your controlling, codependent father that you can only speak with him once a month? Can you finally stop pursuing a close relationship with your abusive sister? Can you scale back on texting that manipulative friend, stop chasing the person giving you the silent treatment, or set better boundaries at work? Limiting toxic interactions is an act of self-love. Imagine what your life would be like if you exclusively spent time with people who built you up instead of tearing you down? *Life-changing.*

PRO TIPS

- Setting boundaries is a form of self-care that fosters self-esteem. Research shows that we tend to respect people more, not less, when they set clear boundaries and enforce them. (This is particularly true of children and their parents.)
- Though you aren't required to offer any explanations, it can be helpful to tell the person how their words and behaviors make you feel when you set your boundary. This strategy can increase the chances your boundaries will be respected. For example: *When you talk to me in that tone of voice it makes me feel disrespected.* Or: *The reason I'm asking you to stop piling on additional demands is because I'm already overwhelmed.*
- It's normal to feel emotions like guilt and shame when you first start setting boundaries. Many of us have been acculturated to please others, particularly family members, often at our own expense. Going against this training can feel bad. Working with a therapist you like and trust can offer valuable support while you're learning to set boundaries, particularly with family members.

IV. Treating Trauma: A Hopeful Path Forward

Because trauma lives in our minds *and* our bodies, its treatment must also target both. Trauma affects our thoughts and emotions; we might perceive the world as dangerous and unsafe, or start to predict that terrible things will happen. Hallie experienced this firsthand after her father threatened her with a gun. We might experience intrusive, unwanted thoughts. Trauma can also result in flashbacks, a reexperiencing of past events so vivid it feels like they're happening all over again. Not surprisingly, trauma is often accompanied by anxiety, panic attacks, anger, and hopelessness. These feelings typically compel us to avoid the people and places that remind us of the trauma. Together, these symptoms keep our brains stuck in a state of alarm.

Physiologically, there are a host of changes in the body that treatment must address, too, from hyperventilation and muscle tension to insomnia and sweaty nightmares. People who have survived a trauma, like Hallie, often feel edgy, alert, hypervigilant, and attuned to small changes in the environment. We might engage in "safety behaviors," like repeatedly checking to ensure doors and windows are locked. Trauma also hijacks our nervous system, launching it into overdrive and exaggerating our five senses. We can become easily overwhelmed by bright lights, loud noises, unexpected touch—and pain.

So, what can we do to help? First and foremost, I recommend getting extremely familiar with the relaxation and self-soothing strategies in Chapter 12, starting with diaphragmatic breathing, guided imagery, and mindfulness. These can improve trauma symptoms by:

- Shutting off the sympathetic fight-or-flight response while activating the parasympathetic rest-and-repair system
- Helping us feel safe and instilling a sense of calm
- Engaging the prefrontal cortex to bring us back to the present moment
- Soothing and regulating the amygdala and associated neural circuits that encode fear and alarm

- Helping us ground and self-soothe
- Reducing pain volume

A multitude of trauma treatments are also now available that target co-occurring trauma and chronic pain, and can significantly reduce symptoms of both.[13,14] A list of these science-backed strategies is included below. I highly recommend trying more than one on your journey to recovery to identify those that work best for you. While you can read through workbooks and treatment manuals on your own, trauma is best treated in the care of a trained provider. The treatments below are typically administered by psychologists, social workers, and other licensed psychotherapists.

- *Seeking Safety (SS).* SS is a treatment protocol frequently used for people struggling with PTSD and simultaneous substance use issues. It teaches fundamental coping and brain-reprocessing skills, has an easy-to-use manual, and can be used with individual patients or groups. As a predoctoral student in graduate school, I ran Seeking Safety groups at a veterans hospital for veterans recently returned from Iraq and Afghanistan living with PTSD and substance use disorders. Consistent with the research, after our ten-week course, the veterans reported less substance use and significantly fewer trauma symptoms.[15]
- *Cognitive processing therapy (CPT).* CPT has evidence of effectiveness for treating the trauma of adverse childhood experiences (ACEs) like child abuse, combat, rape and sexual trauma, and natural disasters, among others. It helps us create new understandings and conceptualizations of past traumas by re-writing and re-telling our story. Doing this in a supportive, calming, controlled environment helps us make new meaning of the experience. It also teaches our brains that trauma symptoms and memories, while difficult and distressing, aren't dangerous, and that we can ultimately handle them.
- *Eye movement desensitization and reprocessing therapy (EMDR).* EMDR is a popular trauma treatment that involves processing

traumatic memories while focusing on other stimuli (visual, auditory, sensory). There's some controversy around whether eye movements and sensory stimuli meaningfully contribute to the treatment's effectiveness, but research studies support its utility.[16]

- *Trauma-focused cognitive behavioral therapy (TF-CBT).* TF-CBT helps us retrain our brains to reduce reactivity to triggers, improve coping, and reduce symptoms.[17] Similar to how CBT-I is tailored for sleep and CBT-CP is tailored for chronic pain, TF-CBT was specifically designed for treating trauma. It addresses traumatic memories and thoughts (e.g., "the world is an unsafe place; bad things are going to happen"), emotions and physiological responses (e.g., anxiety, fear, hypervigilance), and common coping behaviors (e.g., avoidance, substance use, rage responses) to help us resume functioning and feel better.

- *Prolonged exposure (PE).* PE uses the power of exposure therapy to help us desensitize our brain and nervous system to triggers. It includes writing or telling our trauma story in small, gradual doses in a safe environment. As with any good trauma treatment, exposures are done very slowly with a trained provider. PE can help us begin to approach places, situations, and people we've been avoiding to break the trauma-pain cycle.

- *Somatic Experiencing (SE) therapy.* Trauma lives in our bodies, but woefully few treatments directly support our body's recovery. By integrating body-based and cognitive interventions like movement, breathwork, mindfulness, and psychotherapy, SE treats trauma by changing our relationship with physical symptoms and sensations—including pain. It can help soothe our nervous system, help our bodies release tension, and relieve chronic pain.[18]

- *Movement medicine: yoga, dance, exercise.* After a trauma, our stress system is thrown into overdrive. All of that excess energy and activation need a place to go. Experts recommend using physically expressive activities like yoga, dance, and other forms of movement to help release this pent-up energy, eat up stress hormones, and reduce trauma symptoms.[19] While not a panacea on their own, movement medicine can meaningfully reduce PTSD symptomatology, anxiety, and sleep issues, and help the body recalibrate.

- *Medications.* The most recent guidelines for trauma treatment strongly recommend using the therapies listed above *over* medications.[20] Currently, the only FDA-approved medications for trauma are a few selective serotonin reuptake inhibitors (SSRIs). However, scientists have found these aren't particularly effective.[21] And this makes sense: Trauma (like pain!) is a biopsychosocial phenomenon that requires a biopsychosocial solution. As with all medications, my rule of thumb is this: Three months is sufficient time for a drug to do its job. If a medication hasn't meaningfully improved your symptoms in this timeframe, ask your doctor to help you taper off and try a different approach.

PRO TIPS

- For trauma, as for pain, there's no quick fix. Effective treatment requires time, commitment, and work to change the brain and activate the body's pharmacy. The good news is that, if you've survived a trauma, you already have the resilience, persistence, and capacity to get through treatment and reach the other side. I promise that's true. The only way to begin this process is to begin it.
- *You are not alone.* Trauma is best navigated with support. To find healthcare providers trained in trauma treatments like Seeking Safety, trauma-focused CBT, EMDR, and Somatic Experiencing to support you on your journey, try:

- The Trauma Institute (ticti.org), Trauma Research Foundation (TraumaResearchFoundation.org), VA Department of Veterans Affairs (ptsd.va.gov), and the International Society for Trauma Stress Studies (istss.org/portal/find-a-clinician)
- Somatic Experiencing International (TraumaHealing.org)
- Trauma-Focused CBT certification program (tfcbt.org)
- EMDR International Association (emdria.org)
- Seeking Safety, also known as Seeking Strength for veterans (SeekingSafety.org)
- The Seeking Safety workbook, which doubles as a treatment manual, is also available online.

- The goal isn't to find a trauma therapist quickly, but rather to identify a provider you like, trust, and want to work with. Meet them, talk to them, try them on. Then select the best fit for your unique brain.
- Trauma treatments are often offered in group formats, particularly at large healthcare institutions or veterans hospitals. While different formats work for different people, groups are definitely worth trying. In addition to treatment, they also offer a community and peer support. An experienced group leader will establish a safe space, set clear boundaries around triggering language, and be open to questions and concerns.
- Exercise and movement are particularly important if you're dealing with the aftermath of trauma and experiencing symptoms. There are a zillion ways to burn off the excess energy of a stress response in overdrive—including walking, which is completely free. If you're interested in integrating dance and/or yoga into your treatment, try the International Association of Yoga Therapists (iayt.org) or

the American Dance Therapy Association (adta.org). Yoga has become particularly popular in recent years, and chances are high there's a yoga studio near you.

- Healthcare Providers:
    - When treating a patient with chronic pain, *always assess for trauma*. You may be the one person who has linked trauma to pain for this patient, and your referrals can change lives. To assess for ACEs in both children and adults, screening tools are available online (try ACESaware.org). The Childhood Trauma Questionnaire for adults is a well-established measure that retrospectively assesses for childhood maltreatment. A structured clinical interview such as the Clinician-Administered PTSD Scale is considered the gold standard for diagnosing PTSD.
    - It's never too late to become trauma informed. Trauma-informed healthcare is a better, more evolved lens we can use to better help our patients. It shifts our perspective from "What's wrong with you?" to "What happened to you?"[22] For providers who want to become trauma informed, the Trauma-Informed Care Implementation Resource Center (traumainformedcare.chcs.org) is a great resource.

Conclusion

THE FUTURE OF PAIN MEDICINE IS YOU

Pain, despite its invisibility, has a face. It just isn't the face smiling at us from those 0-to-10 pain scales. Instead, it's the faces of the real people behind them. I wrote this book because of those faces. I was tired of seeing my patients, many of them children, languishing for years, doomed to deteriorate, given pills for their pain and nothing more.

My patients, by and large, get better: They get out of bed and back to life, their pain faded to background noise. This isn't magic. It's because the treatment for pain, the *true* treatment, requires looking beyond the diagnosis to the person beneath.

As a teenager and into adulthood, I read every Oliver Sacks book I could get my eager hands on. A neurologist and acclaimed science writer, Dr. Sacks was famous for many things, one of which was asking his patients nontraditional questions. He examined not just their brains and bodies, but also their environments, visiting them at home and work. He examined their hobbies, families, and support networks. This, he was told, was a liability—an inefficient waste of time. But it was actually a gift.

His patients, he recognized, came with context, and that context helped *him* help *them*. Seeing his patients as whole people—complex human beings with textured lives, histories, and backgrounds—changed both the diagnosis and the prescription. His stories ultimately speak to us because we know, intuitively, that this whole-person approach to medicine isn't just the best approach: It's the *only* one.

People living with chronic pain become "incurable" not because their chronic pain can't be cured—but because we aren't going wide enough, and we aren't going deep enough. Modern healthcare isn't currently structured to support this kind of thorough examination, in part because insurance companies don't reimburse them. In turn, the clinics and hospital systems that employ our healthcare providers don't allot sufficient time.

But the pain revolution is here, and the time is now. There is no reason for us to continue treating pain with an outdated, ineffective tool kit. As providers, it isn't too late to change the way we practice medicine—to change the language we use, the solutions we recommend, and the prescriptions we write. To spend a few minutes getting to know the struggling single mother beneath that gastrointestinal distress, too afraid to take the bus home. To learn more about the lonely engineer attached to that Fabry diagnosis, certain he'll never walk again. To finally stop stigmatizing the suffering of a teenager, and acknowledge that his pain is real.

As patients and people who experience pain, it isn't too late for us to assess our own health on a broader scale—to examine the foods we eat, how well we sleep, and our trauma histories. To reconnect physical and emotional pain, the body and the brain. *To treat the person instead of the body part.* Because you, like Dr. Sacks, like me, are also a pain detective.

We can start constructing pain recipes, adopt "biopsychosocial" and "Pain Dial" language in our homes and hospitals, and incorporate tools like pacing, mindfulness, and social medicine. We can minimize danger messages, replacing them with heaping doses of hope. We can even lobby insurance companies until treatments like biofeedback and CBT are as reimbursable as that second surgery.

When you put down this book and walk away, our work here will not be done. For the foreseeable future, the world around you will attempt to undo everything you've just learned. You will be told, repeatedly, that your pain is purely biomedical and requires a biomedical solution—undoubtedly in the form of pills or procedures. Or you'll be told your pain is all in your head. This means that you have one crucial job ahead of you, and it is this: To remain rooted in the Truth. To remember that there are *many* ways to change your brain and activate your body's pharmacy beyond those offered at the drugstore and in the operating room.

Margaret Mead, anthropologist and author, once said that we should never doubt that a small group of thoughtful, committed citizens can change the world, because it's the only thing that ever has. So, hear me now: We need not wait for change to come from the top down. We need not wait for our profit-driven healthcare system to reform, or for the antiquated biomedical model to join the twenty-first century. In fact, we need not wait another second to transform the way we understand and treat pain.

Because *you* are the change. *I* am the change.

We are the faces of pain, and together, *we* are the change.

What kind of pain detective will you decide to be…?

Acknowledgments

To the entire team at Hachette Grand Central, Karyn and Colin in particular: Thank you for taking a chance on me and on this movement, and for getting this book into the hands of the people who need it most. To my husband: Thank you for the incredible support, endless patience, generous love, nature breaks, tech support, beautiful meals, reference managing, science articles read in Snoop Dogg's voice, and incomparable calm. To my mom: Thank you for believing me when I told you at age six that I'd be a writer one day, and for the endless words of encouragement. Thanks to Michèle and Normand for advice and endless cheerleading.

To my team: Thank you to editorial consultants Jenna Free for thought-partnership, countless re-orgs, and Seattle brainstorms, and Jeff Alexander for thoughtful edits; the talented Eva Huzella for gorgeous illustrations; Pamela Ash for permissions help; Pete Garceau for kismet cover collaboration, and Amanda Silver for the impossible meet-cute. (For those who have asked: No, this book was not ghostwritten.) Thanks to Debbie Berne, Allison, Kevin, Janey, Sam, Shelly, Molly, Myles, Ida, and Camille for invaluable input on titles and covers. Thanks to former agent Dr. Ross for brokering an unheard-of fifteen-publisher auction, Eric Rayman and Carolyn Levin for guidance every step of the way, Mason Muñoz for last-minute endnote support, Allison Light and Kat Zheleznyak for next-level consultation and friendship, Eva Huzella and Janey Keig for design consults, and assistants Nicole and Mason for helping me stay organized.

Thank you to Dr. Jack Stern, brilliant neurosurgeon, author, and friend, for his beautiful foreword, sage consults, and invaluable feedback and support. To Drs. John Loeser and Joel Katz, pioneering pain scientists and fellow revolutionaries, for fact-checks and feedback on initial drafts. To Drs. Tor Wager, Fadel Zeidan, Erik Peper, and Michael Gold for generous consults.

To Brown University's Dr. J. Michael Walker, chair of my honors thesis on pain neuroscience; Dr. Mark Bear, neuroscientist and friend, who taught Neuro 1 and inadvertently changed my career path; and Dean of Biology Marjorie Thompson, in memory and with love, for mentorship, guidance, and friendship.

Thank you to every friend who offered words of encouragement and evenings off. To Victor, Marie-Helene, and Joyce for use of their peaceful cabins in the woods. To my coworking companions at Page Street and the Notto for company and camaraderie, Helen Gaughran in particular. And finally, to my patients, who continue to teach and guide me every day: Thank you for letting me into your lives, and for permitting me to share facets of your important stories. Your voices matter—keep lifting them!

Endnotes

This is a partial reference list; an extended list of citations and sources can be found on Zoffness.com and http://grandcentralpublishing.com/. Some of the activities in part III, chapters 10–14, have been adapted from Zoffness, R. (2020). *The Pain Management Workbook: Powerful CBT and Mindfulness Skills to Take Control of Pain and Reclaim Your Life.* New Harbinger Publications.

1. EVERYTHING YOU'VE BEEN TOLD ABOUT PAIN IS WRONG

1. Institute of Medicine Committee on Advancing Pain Research, Care, and Education (2011). *Relieving Pain in America: A Blueprint for Transforming Prevention, Care, Education, and Research.* National Academies Press.
2. Lucas, J., & Sohi, I. (2024). *Chronic Pain and High-Impact Chronic Pain in US Adults, 2023.* U.S. Department of Health & Human Services, Centers for Disease Control and Prevention, National Center for Health Statistics.
3. Mannion, A., Brox, J., & Fairbank, J. (2016). Consensus at last! Long-term results of all randomized controlled trials show that fusion is no better than non-operative care in improving pain and disability in chronic low back pain. *Spine Journal, 16*(5), 588–590.
4. Traeger, A., Gilbert, S., Harris, I., et al. (2023). Spinal cord stimulation for low back pain. *Cochrane Database of Systematic Reviews*, (3).
5. Shipton, E., Bate, F., Garrick, R., et al. (2018). Systematic review of pain medicine content, teaching, and assessment in medical school curricula internationally. *Pain and Therapy, 7*, 139–161.
6. Thompson, K., Johnson, M. I., Milligan, J., et al. (2018). Twenty-five years of pain education research—what have we learned? Findings from a comprehensive scoping review of research into pre-registration pain education for health professionals. *Pain, 159*(11), 2146–2158.

7   Holst, J. (2022). Biomedical Perspective: Critical Assessment of an Outdated Concept. *Open Access J Biomed Sci, 4*(2), 000435.

8   Matthews, J., Zoffness, R., & Becker, D. (2021). Integrative pediatric pain management: Impact & implications of a novel interdisciplinary curriculum. *Complementary Therapies in Medicine, 59*, 102721.

9   Loeser, J. (2015). The education of pain physicians. *Pain Medicine, 16*(2), 225–229.

10 Bendelow, G. (2013). Chronic pain patients and the biomedical model of pain. *AMA Journal of Ethics, 15*(5), 455–459.

11 Muhuri, P., Gfroerer, J., & Davies, M. C. (2013). CBHSQ data review. *Center for Behavioral Health Statistics and Quality, SAMHSA, 1*, 17.

12 Fillingim, M., Tanguay-Sabourin, C., Parisien, M., et al. (2025). Biological markers and psychosocial factors predict chronic pain conditions. *Nature Human Behaviour*, 1–16.

13 Adams, L., & Turk, D. (2018). Central sensitization and the biopsychosocial approach to understanding pain. *Journal of Applied Biobehavioral Research, 23*(2), e12125.

14 Martucci, K., & Mackey, S. (2016). Imaging pain. *Anesthesiology Clinics, 34*(2), 255–269.

15 Garland, E. (2012). Pain processing in the human nervous system: A selective review of nociceptive and biobehavioral pathways. *Primary Care: Clinics in Office Practice, 39*(3), 561–571.

16 Wade, D., & Halligan, P. (2017). The biopsychosocial model of illness: A model whose time has come. *Clinical Rehabilitation, 31*(8), 995–1004.

17 Bevers, K., Watts, L., Kishino, N. D., et al. (2016). The biopsychosocial model of the assessment, prevention, and treatment of chronic pain. *U.S. Neurol, 12*(2), 98–104.

18 Tick, H., Nielsen, A., Pelletier, K., et al. (2018). Evidence-based nonpharmacologic strategies for comprehensive pain care: The consortium pain task force white paper. *Explore, 14*(3), 177–211.

19 Bujak, B., Regan, E., Beattie, P., et al. (2019). The effectiveness of interdisciplinary intensive outpatient programs in a population with diverse chronic pain conditions: A systematic review and meta-analysis. *Pain Management, 9*(4), 417–429.

20 Elbers, S., Wittink, H., Konings, S., et al. (2022). Longitudinal outcome evaluations of Interdisciplinary Multimodal Pain Treatment programmes for patients with chronic primary musculoskeletal pain: A systematic review and meta-analysis. *European Journal of Pain, 26*(2), 310–335.

21 Andronis, L., Kinghorn, P., Qiao, S., et al. (2017). Cost-effectiveness of non-invasive and non-pharmacological interventions for low back pain: A systematic literature review. *Applied Health Economics and Health Policy*, *15*, 173–201.

22 Dowell D., Ragan K., Jones C., et al. (2022). CDC Clinical Practice Guideline for Prescribing Opioids for Pain—United States. *MMWR Recommendations & Reports*, *71*(No. RR-3), 1–95.

2. THE REAL SCIENCE OF PAIN

1 Howard, D. (2011). Bolt: "I want to do wild things." *ESPN, the Magazine*. Nov. 29, 2011.

2 Gijy, J. (2021). Olympic Legend Usain Bolt Admits He Never Corrected His Scoliosis Throughout Career. *Essentially Sports*. Sept. 6, 2021.

3 Longman, J. (2017). Something Strange in Usain Bolt's Stride: Bolt is the fastest sprinter ever in spite of—or because of?—an uneven stride that upends conventional wisdom. *New York Times*. July 20, 2017.

4 Brosh, A. (2010). Boyfriend Doesn't Have Ebola. Probably. *Hyperbole and a Half*. Feb 10, 2010. https://hyperboleandahalf.blogspot.com/2010/02/boyfriend-doesnt-have-ebola-probably.html.

5 Raja, S., Carr, D., Cohen, M., et al. (2020). The revised International Association for the Study of Pain definition of pain: Concepts, challenges, and compromises. *Pain*, *161*(9), 1976–1982.

6 Van Der Miesen, M., Lindquist, M., & Wager, T. (2019). Neuroimaging-based biomarkers for pain: State of the field and current directions. *Pain reports*, *4*(4), e751.

7 Mercer L., Chen, C., Gilam, G., et al. (2021). Brain circuits for pain and its treatment. *Science translational medicine*, *13*(619), eabj7360.

8 Martucci K., & Mackey, S. (2018). Neuroimaging of pain: Human evidence and clinical relevance of central nervous system processes and modulation. *Anesthesiology*, *128*(6), 1241–1254.

9 Garcia- Larrea L., Bastuji H. (2018) Pain and consciousness. *Progress in Neuro-Psychopharmacology & Biological Psychiatry*. 87:193–199.

10 Moseley, G. (2007). Reconceptualising pain according to modern pain science. *Physical Therapy Reviews*, *12*(3), 169–178.

11 Eveleth, R. (2014). Americans Are More into BDSM Than the Rest of the World. *Smithsonian Magazine* (Feb. 10, 2014).

12 Leknes, S., Berna, C., Lee, M., et al. (2013). The importance of context: When relative relief renders pain pleasant. *Pain*, *154*(3), 402–410.

13 Carlino, E., & Benedetti, F. (2016). Different contexts, different pains, different experiences. *Neuroscience, 338*, 19–26.

14 Dimsdale, J., & Dantzer, R. (2007). A biological substrate for somatoform disorders: Importance of pathophysiology. *Psychosomatic Medicine, 69*(9), 850–854.

15 Fisher J., Hassan D., O'Connor N. (1995). Minerva. *British Medical Journal.* 310, 70.

16 Gartner, E. (2005). Cause of "toothache" was 4-inch nail in man's skull. *Seattle Times*, Jan 17, 2005.

17 Kasch, R., Truthmann, J., Hancock, M., et al. (2022). Association of lumbar MRI findings with current and future back pain in a population-based cohort study. *Spine, 47*(3), 201–211.

18 Deyo, R., Mirza, S., Turner, J., et al. (2009). Overtreating chronic back pain: Time to back off? *Journal of the American Board of Family Medicine, 22*(1), 62–68.

19 Brinjikji, W., Luetmer, P., Comstock, B., et al. (2015). Systematic literature review of imaging features of spinal degeneration in asymptomatic populations. *American Journal of Neuroradiology, 36*(4), 811–816.

20 Nakashima, H., Yukawa, Y., Suda, K., et al. (2015). Abnormal findings on magnetic resonance images of the cervical spines in 1211 asymptomatic subjects. *Spine, 40*(6), 392–398.

21 Harris, I. A., Sidhu, V., Mittal, R., et al. (2020). Surgery for chronic musculoskeletal pain: The question of evidence. *Pain, 161*, S95–S103.

22 Bertozzi, L., Negrini, S., Agosto, D., et al. (2021). Posture and time spent using a smartphone are not correlated with neck pain and disability in young adults: A cross-sectional study. *Journal of Bodywork and Movement Therapies, 26*, 220–226.

23 Moayedi, M., & Davis, K. (2013). Theories of pain: From specificity to gate control. *Journal of Neurophysiology, 109*(1), 5–12.

3. IF THE BRAIN CAN CHANGE, PAIN CAN CHANGE

1 Kuner, R., & Flor, H. (2017). Structural plasticity and reorganisation in chronic pain. *Nature Reviews Neuroscience, 18*(1), 20–30.

2 Basbaum, A., Bautista, D., Scherrer, G., et al. (2009). Cellular and molecular mechanisms of pain. *Cell, 139*(2), 267–284.

3 Woolf, C. J. (2011). Central sensitization: Implications for the diagnosis and treatment of pain. *Pain, 152*(3), S2–S15.

4   Moseley, G., & Flor, H. (2012). Targeting cortical representations in the treatment of chronic pain: A review. *Neurorehabilitation and Neural Repair*, *26*(6), 646–652.

5   Nijs, J., George, S., Clauw, D., et al. (2021). Central sensitisation in chronic pain conditions: Latest discoveries and their potential for precision medicine. *The Lancet Rheumatology*, *3*(5), e383–e392.

6   Latremoliere, A., & Woolf, C. (2009). Central sensitization: A generator of pain hypersensitivity by central neural plasticity. *Journal of Pain*, *10*(9), 895–926.

7   Gold, M., & Gebhart, G. (2010). Nociceptor sensitization in pain pathogenesis. *Nature Medicine*, *16*(11), 1248–1257.

8   Moseley, G., & Vlaeyen, J. (2015). Beyond nociception: The imprecision hypothesis of chronic pain. *Pain*, *156*(1), 35–38.

9   Crofford, L. (2015). Chronic pain: Where the body meets the brain. *Transactions of the American Clinical and Climatological Association*, *126*, 167.

10 Fine, P. (2011). Long-term consequences of chronic pain: Mounting evidence for pain as a neurological disease and parallels with other chronic disease states. *Pain Medicine*, *12*(7), 996–1004.

11 Katz, J., & Seltzer, Z. (2009). Transition from acute to chronic postsurgical pain: Risk factors and protective factors. *Expert Review of Neurotherapeutics*, *9*(5), 723–744.

12 Price, T., & Dussor, G. (2014). Evolution: The advantage of 'maladaptive' pain plasticity. *Current Biology*, *24*(10), R384–R386.

13 McCarberg, B., & Peppin, J. (2019). Pain pathways and nervous system plasticity: Learning and memory in pain. *Pain Medicine*, *20*(12), 2421–2437.

14 Volcheck, M., Graham, S., Fleming, K., et al. (2023). Central sensitization, chronic pain, and other symptoms: Better understanding, better management. *Cleveland Clinic Journal of Medicine*, *90*(4), 245–254.

15 Garland, E. (2012). Pain processing in the human nervous system: A selective review of nociceptive and biobehavioral pathways. *Primary Care: Clinics in Office Practice*, *39*(3), 561–571.

16 Ji, R., Nackley, A., Huh, Y., et al. (2018). Neuroinflammation and central sensitization in chronic and widespread pain. *Anesthesiology*, *129*(2), 343.

17 Volcheck, M., Graham, S., Fleming, K., et al. (2023). Central sensitization, chronic pain, and other symptoms: Better understanding, better management. *Cleveland Clinic Journal of Medicine*, *90*(4), 245–254.

18 Raffaeli, W., & Arnaudo, E. (2017). Pain as a disease: An overview. *Journal of Pain Research*, 2003–2008.

19 Volcheck, M., Graham, S., Fleming, K., et al. (2023). Central sensitization, chronic pain, and other symptoms: Better understanding, better management. *Cleveland Clinic Journal of Medicine*, *90*(4), 245–254.

20 Aron, E., Aron, A., & Jagiellowicz, J. (2012). Sensory processing sensitivity: A review in the light of the evolution of biological responsivity. *Personality and Social Psychology Review*, *16*(3), 262–282.

21 Morellini, L., Izzo, A., Celeghin, A., et al. (2023). Sensory processing sensitivity and social pain: A hypothesis and theory. *Frontiers in Human Neuroscience*, *17*, 1135440.

22 Clark, J., Nijs, J., Yeowell, G., et al. (2019). Trait sensitivity, anxiety, and personality are predictive of central sensitization symptoms in patients with chronic low back pain. *Pain Practice*, *19*(8), 800–810.

23 Lotze, M., & Moseley, G. (2015). Theoretical considerations for chronic pain rehabilitation. *Physical Therapy*, *95*(9), 1316–1320.

24 Greenwald, J., & Shafritz, K. (2018). An integrative neuroscience framework for the treatment of chronic pain: From cellular alterations to behavior. *Frontiers in Integrative Neuroscience*, *12*, 18.

25 Gigandet, M. (2024). The Extra Mile. *Atavist Magazine*. June 2024.

5. PAIN IS EMOTIONAL

1 Lumley, M., Cohen, J., Borszcz, G., et al. (2011). Pain and emotion: A biopsychosocial review of recent research. *Journal of Clinical Psychology*, *67*(9), 942–968.

2 Burns, J., Quartana, P., Gilliam, W., et al. (2008). Effects of anger suppression on pain severity and pain behaviors among chronic pain patients: Evaluation of an ironic process model. *Health Psychology*, *27*(5), 645.

3 Martucci, K., & Mackey, S. (2018). Neuroimaging of pain: Human evidence and clinical relevance of central nervous system processes and modulation. *Anesthesiology*, *128*(6), 1241.

4 Vastag, B. (2003). Scientists find connections in the brain between physical and emotional pain. *JAMA*, *290*(18), 2389–2390.

5 Gilam, G., Gross, J., Wager, T., et al. (2020). What is the relationship between pain and emotion? Bridging constructs and communities. *Neuron*, *107*(1), 17–21.

6 Villemure, C., & Schweinhardt, P. (2010). Supraspinal pain processing: Distinct roles of emotion and attention. *Neuroscientist*, *16*(3), 276–284.

7 Wiech, K., & Tracey, I. (2009). The influence of negative emotions on pain: Behavioral effects and neural mechanisms. *Neuroimage*, *47*(3), 987–994.

8   Roy, M., Piché, M., Chen, J., et al. (2009). Cerebral and spinal modulation of pain by emotions. *Proceedings of the National Academy of Sciences*, *106*(49), 20900–20905.

9   Bushnell, M., Čeko, M., & Low, L. (2013). Cognitive and emotional control of pain and its disruption in chronic pain. *Nature Reviews Neuroscience*, *14*(7), 502–511.

10 Finan, P., & Garland, E. (2015). The role of positive affect in pain and its treatment. *Clinical Journal of Pain*, *31*(2), 177–187.

11 Hitchcock, E., Hassett, A., & Wager, T. (2019). Effects of positive emotion on pain: Mechanisms and interventions. In J. Gruber (Ed.), *The Oxford Handbook of Positive Emotion and Psychopathology* (pp. 444–452). Oxford University Press.

12 Dunbar, R., Baron, R., Frangou, A., et al. (2012). Social laughter is correlated with an elevated pain threshold. *Proceedings of the Royal Society B: Biological Sciences*, *279*(1731), 1161–1167.

13 Mason, I. (2013) Laughing Pain Away. *Medical News Today*. October 15, 2013.

14 Lopes-Júnior, L., Bomfim, E., Olson, K., et al. (2020). Effectiveness of hospital clowns for symptom management in paediatrics: Systematic review of randomised and non-randomised controlled trials. *BMJ*, *371*, m4290.

15 Tse, M., Lo, A., Cheng, T., et al. (2010). Humor therapy: Relieving chronic pain and enhancing happiness for older adults. *Journal of Aging Research*, *2010*(1), 343574.

16 Cross, M., Acevedo, A., Leger, K., et al. (2023). How and why could smiling influence physical health? A conceptual review. *Health Psychology Review*, *17*(2), 321–343.

17 Pérez-Aranda, A., Hofmann, J., Feliu-Soler, A., et al. (2019). Laughing away the pain: A narrative review of humour, sense of humour and pain. *European Journal of Pain*, *23*(2), 220–233.

18 Lapierre, S., Baker, B., & Tanaka, H. (2019). Effects of mirthful laughter on pain tolerance: A randomized controlled investigation. *Journal of Bodywork and Movement Therapies*, *23*(4), 733–738.

19 Abdallah, C., & Geha, P. (2017). Chronic pain and chronic stress: Two sides of the same coin? *Chronic Stress*, *1*.

20 Jennings, E., Okine, B., Roche, M., et al. (2014). Stress-induced hyperalgesia. *Progress in Neurobiology*, *121*, 1–18.

21 Bomholt, S., Harbuz, M., Blackburn-Munro, G., et al. (2004). Involvement and role of the hypothalamo-pituitary-adrenal (HPA) stress axis in animal models of chronic pain and inflammation. *Stress*, *7*(1), 1–14.

22 Hannibal, K., & Bishop, M. (2014). Chronic stress, cortisol dysfunction, and pain: A psychoneuroendocrine rationale for stress management in pain rehabilitation. *Physical Therapy, 94*(12), 1816–1825.

23 Lunde, C., & Sieberg, C. (2020). Walking the tightrope: A proposed model of chronic pain and stress. *Frontiers in Neuroscience, 14*, 270.

24 Channel Action 7 News New Mexico, KOAT.com (2012, July 17). Man lifts car off 3-year-old child [Video]. YouTube. https://www.youtube.com/watch?v=J-qQbIiOtYM.

25 Yilmaz, P., Diers, M., Diener, S., et al. (2010). Brain correlates of stress-induced analgesia. *Pain, 151*(2), 522–529.

26 De La Rosa, J., Brady, B., Ibrahim, M., et al. (2024). Co-occurrence of chronic pain and anxiety/depression symptoms in US adults: Prevalence, functional impacts, and opportunities. *Pain, 165*(3), 666–673.

27 Strobel, C., Hunt, S., Sullivan, R., et al. (2014). Emotional regulation of pain: The role of noradrenaline in the amygdala. *Science China Life Sciences, 57*, 384–390.

28 Burston, J., Valdes, A., Woodhams, S., et al. (2019). The impact of anxiety on chronic musculoskeletal pain and the role of astrocyte activation. *Pain, 160*(3), 658–669.

29 Neugebauer, V. (2015). Amygdala pain mechanisms. *Pain Control*, 261–284.

30 IsHak, W., Wen, R., Naghdechi, L., et al. (2018). Pain and depression: A systematic review. *Harvard Review of Psychiatry, 26*(6), 352–363.

31 Tang, N., & Crane, C. (2006). Suicidality in chronic pain: A review of the prevalence, risk factors and psychological links. *Psychological Medicine, 36*(5), 575–586.

32 Hooten, W. (2016). Chronic pain and mental health disorders: Shared neural mechanisms, epidemiology, and treatment. *Mayo Clinic Proceedings, 91*(7), 955–970.

33 Sheng, J., Liu, S., Wang, Y., et al. (2017). The link between depression and chronic pain: Neural mechanisms in the brain. *Neural Plasticity, 2017*(1), 9724371.

6. PAIN IS COGNITIVE

1 Kiecolt-Glaser, J., McGuire, L., Robles, T., et al. (2002). Psychoneuroimmunology: Psychological influences on immune function and health. *Journal of Consulting and Clinical Psychology, 70*(3), 537.

2 Quartana, P., Campbell, C., & Edwards, R. (2009). Pain catastrophizing: A critical review. *Expert Review of Neurotherapeutics, 9*(5), 745–758.

3 Arntz, A., & Claassens, L. (2004). The meaning of pain influences its experienced intensity. *Pain, 109*(1–2), 20–25.

4   Atlas, L & Wager, T. (2012). How expectations shape pain. *Neuroscience Letters, 520*(2), 140–148.

5   González Aroca, J., Díaz, Á., Navarrete, C., et al. (2023). Fear-avoidance beliefs are associated with pain intensity and shoulder disability in adults with chronic shoulder pain: A cross-sectional study. *Journal of Clinical Medicine, 12*(10), 3376.

6   Sullivan, M., Bishop, S., & Pivik, J. (1995). The pain catastrophizing scale: Development and validation. *Psychological Assessment, 7*(4), 524.

7   Simic, K., Savic, B., & Knezevic, N. (2024). Pain catastrophizing: How far have we come. *Neurology International, 16*(3), 483–501.

8   Galambos, A., Szabó, E., Nagy, Z., et al. (2019). A systematic review of structural and functional MRI studies on pain catastrophizing. *Journal of Pain Research, 12*, 1155–1178.

9   Richter, C. (1957). On the phenomenon of sudden death in animals and man. In *Psychopathology: A Source Book* (pp. 234–242). Harvard University Press.

10 Rozanski, A., Bavishi, C., Kubzansky, L., et al. (2019). Association of optimism with cardiovascular events and all-cause mortality: A systematic review and meta-analysis. *JAMA Network Open, 2*(9), e1912200.

11 Hanssen, M., Peters, M., Vlaeyen, J., et al. (2013). Optimism lowers pain: Evidence of the causal status and underlying mechanisms. *Pain, 154*(1), 53–58.

12 Forte, A., Guliyeva, G., McLeod, H., et al. (2022). The impact of optimism on cancer-related and postsurgical cancer pain: A systematic review. *Journal of Pain and Symptom Management, 63*(2), e203–e211.

13 Basten-Günther, J., Peters, M., & Lautenbacher, S. (2019). Optimism and the experience of pain: A systematic review. *Behavioral Medicine, 45*(4), 323–339.

14 Simons, D., & Chabris, C. (1999). Gorillas in our midst: Sustained inattentional blindness for dynamic events. *Perception, 28*(9), 1059–1074.

15 Bushnell, M., Čeko, M., & Low, L. (2013). Cognitive and emotional control of pain and its disruption in chronic pain. *Nature Reviews Neuroscience, 14*(7), 502–511.

16 Torta, D., Legrain, V., Mouraux, A., et al. (2017). Attention to pain! A neurocognitive perspective on attentional modulation of pain in neuroimaging studies. *Cortex, 89*, 120–134.

17 Petersen, S., & Posner, M. (2012). The attention system of the human brain: 20 years after. *Annual Review of Neuroscience, 35*(1), 73–89.

18 Wiech, K. (2016). Deconstructing the sensation of pain: The influence of cognitive processes on pain perception. *Science*, 354(6312), 584–587.

19 Subnis, U., Starkweather, A., & Menzies, V. (2016). A current review of distraction-based interventions for chronic pain management. *European Journal of Integrative Medicine*, 8(5), 715–722.

20 Baker, N., Polhemus, A., Ospina, E., et al. (2022). The state of science in the use of virtual reality in the treatment of acute and chronic pain: A systematic scoping review. *Clinical Journal of Pain*, 38(6), 424–441.

21 Rischer, K., González-Roldán, A., Montoya, P., et al. (2020). Distraction from pain: The role of selective attention and pain catastrophizing. *European Journal of Pain*, 24(10), 1880–1891.

22 Villemure, C., & Bushnell, M. C. (2002). Cognitive modulation of pain: How do attention and emotion influence pain processing? *Pain*, 95(3), 195–199.

7. PAIN IS SOCIAL

1 Calloway, K. (2019). I spent 16 months in solitary confinement and now I'm fighting to end it. ACLU. July 3, 2019.

2 Eisenberger, N. (2012). The neural bases of social pain: Evidence for shared representations with physical pain. *Psychosomatic Medicine*, 74(2), 126–135.

3 Kross, E., Berman, M., Mischel, W., et al. (2011). Social rejection shares somatosensory representations with physical pain. *Proceedings of the National Academy of Sciences*, 108(15), 6270–6275.

4 DeWall, C., MacDonald, G., Webster, G., et al. (2010). Acetaminophen reduces social pain: Behavioral and neural evidence. *Psychological Science*, 21(7), 931–937.

5 Trang, T., Al-Hasani, R., Salvemini, D., et al. (2015). Pain and poppies: The good, the bad, and the ugly of opioid analgesics. *Journal of Neuroscience*, 35(41), 13879–13888.

6 National Academies of Sciences, Engineering, and Medicine. (2020). *Social Isolation and Loneliness in Older Adults: Opportunities for the Health Care System*. National Academies Press.

7 Cacioppo, J., Cacioppo, S., Capitanio, J., et al. (2015). The neuroendocrinology of social isolation. *Annual Review of Psychology*, 66(1), 733–767.

8 Office of the U.S. Surgeon General (2023). *Our Epidemic of Loneliness and Isolation: The U.S. Surgeon General's Advisory on the Healing Effects of Social Connection and Community*.

9 Wolf, L., & Davis, M. (2014) Loneliness, daily pain, and perceptions of interpersonal events in adults with fibromyalgia. *Health Psychology: Official Journal of the Division of Health Psychology, APA. 33*(9), 929–937.

10 Loeffler, A., & Steptoe, A. (2021). Bidirectional longitudinal associations between loneliness and pain, and the role of inflammation. *Pain, 162*(3), 930–937.

11 Holt-Lunstad, J., Smith, T., Baker, M., et al. (2015). Loneliness and social isolation as risk factors for mortality: A meta-analytic review. *Perspectives on Psychological Science, 10*(2), 227–237.

12 Petitte, T., Mallow, J., Barnes, E., et al. (2015). A systematic review of loneliness and common chronic physical conditions in adults. *Open Psychology Journal, 8*(Suppl 2), 113.

13 Holt-Lunstad, J., Smith, T., & Layton, J. (2010). Social relationships and mortality risk: A meta-analytic review. *PLoS medicine, 7*(7), e1000316.

14 Loggia, M., Mogil, J., & Bushnell, M. (2008). Empathy hurts: Compassion for another increases both sensory and affective components of pain perception. *Pain, 136*(1–2), 168–176.

15 Zaki, J., Wager, T., Singer, T., et al. (2016). The anatomy of suffering: Understanding the relationship between nociceptive and empathic pain. *Trends in Cognitive Sciences, 20*(4), 249–259.

16 Smith, M., Asada, N., & Malenka, R. (2021). Anterior cingulate inputs to nucleus accumbens control the social transfer of pain and analgesia. *Science, 371*(6525), 153–159.

17 Allen, S., Gilbody, S., Atkin, K., et al. (2020). The associations between loneliness, social exclusion and pain in the general population: A N=502,528 cross-sectional UK Biobank study. *Journal of Psychiatric Research, 130*, 68–74.

18 Lieberman, M., & Eisenberger, N. (2006). A pain by any other name (rejection, exclusion, ostracism) still hurts the same: The role of dorsal anterior cingulate cortex in social and physical pain. *Social Neuroscience: People Thinking About Thinking People*, 167–187.

19 Williams, K. D. (2002). *Ostracism: The Power of Silence*. Guilford Press.

20 Williams, K. D. (2011). The pain of exclusion. *Scientific American Mind, 21*(6), 30–37.

21 Hobson, J., Moody, M., Sorge, R., et al. (2022). The neurobiology of social stress resulting from Racism: Implications for pain disparities among racialized minorities. *Neurobiology of Pain, 12*, 100101.

22 Eisenberger, N., & Lieberman, M. (2013). Why it hurts to be left out: The neurocognitive overlap between physical and social pain. In *The Social Outcast* (pp. 109–127). Psychology Press.

23 Holt-Lunstad, J., Smith, T., & Layton, J. (2010). Social relationships and mortality risk: A meta-analytic review. *PLoSMedicine*, *7*(7), e1000316.

24 Field, T., Schanberg, S., Scafidi, F., et al. (1986). Tactile/kinesthetic stimulation effects on preterm neonates. *Pediatrics*, *77*(5), 654–658.

25 Ardiel, E., & Rankin, C. (2010). The importance of touch in development. *Paediatrics & Child Health*, *15*(3), 153–156.

26 Coan, J., Schaefer, H., & Davidson, R. (2006). Lending a hand: Social regulation of the neural response to threat. *Psychological Science*, *17*(12), 1032–1039.

27 Dunbar, R. I. (2010). The social role of touch in humans and primates: Behavioural function and neurobiological mechanisms. *Neuroscience & Biobehavioral Reviews*, *34*(2), 260–268.

28 Ellingsen, D. M., Leknes, S., Løseth, G., et al. (2016). The neurobiology shaping affective touch: Expectation, motivation, and meaning in the multisensory context. *Frontiers in Psychology*, *6*, 1986.

8. PAIN IS ENVIRONMENTAL

1 Craig, K. (2015). Social communication model of pain. *Pain*, *156*(7), 1198–1199.

2 Peacock, S., & Patel, S. (2008). Cultural influences on pain. *Reviews in Pain*, *1*(2), 6–9.

3 Gardner, S. (2022). Serena Williams describes near-death experience she had after giving birth to daughter Olympia. *USA Today*. April 7, 2022.

4 Hoffman, K., Trawalter, S., Axt, J., et al. (2016). Racial bias in pain assessment and treatment recommendations, and false beliefs about biological differences between blacks and whites. *Proceedings of the National Academy of Sciences*, *113*(16), 4296–4301.

5 Schoenthaler, A., & Williams, N. (2022). Looking beneath the surface: Racial bias in the treatment and management of pain. *JAMA Network Open*, *5*(6), e2216281–e2216281.

6 Staton, L., Panda, M., Chen, I., et al. (2007). When race matters: Disagreement in pain perception between patients and their physicians in primary care. *Journal of the National Medical Association*, *99*(5), 532.

7 Goyal M., Kuppermann N., Cleary S., et al. (2015). Racial disparities in pain management of children with appendicitis in emergency departments. *JAMA Pediatrics*, *169*(11), 996–1002.

8 Grol-Prokopczyk, H. (2017). Sociodemographic disparities in chronic pain, based on 12-year longitudinal data. *Pain*, *158*(2), 313–322.

9 Zajacova, A., Grol-Prokopczyk, H., & Zimmer, Z. (2021). Pain trends

among American adults, 2002–2018: Patterns, disparities, and correlates. *Demography*, *58*(2), 711–738.

10 Nanavaty N., Walsh K., Boring B., et al. (2023). Acute ostracism-related pain sensitization in the context of accumulated lifetime experiences of ostracism. *Journal of Pain*, *24*(7), 1229–1239.

11 Hobson, J., Moody, M., Sorge, R., et al. (2022). The neurobiology of social stress resulting from Racism: Implications for pain disparities among racialized minorities. *Neurobiology of Pain*, *12*, 100101.

12 Casale, R., Atzeni, F., Bazzichi, L., et al. (2021). Pain in women: A perspective review on a relevant clinical issue that deserves prioritization. *Pain and Therapy*, *10*, 287–314.

13 Mogil, J. (2012). Sex differences in pain and pain inhibition: Multiple explanations of a controversial phenomenon. *Nature Reviews Neuroscience*, *13*(12), 859–866.

14 Neighmond, P. (2019). Women may be more adept than men at discerning pain. NPR. Aug. 26, 2019.

15 Mazure, C., & Jones, D. (2015). Twenty years and still counting: Including women as participants and studying sex and gender in biomedical research. *BMC Women's Health*, *15*, 1–16.

16 Osborne, N., & Davis, K. (2022). Sex and gender differences in pain. *International Review of Neurobiology*, *164*, 277–307.

17 Lu, H., Hatfield, L., Al-Azazi, S., et al. (2024). Sex-based disparities in acute myocardial infarction treatment patterns and outcomes in older adults hospitalized across 6 high-income countries: An analysis from the International Health Systems Research Collaborative. *Circulation: Cardiovascular Quality and Outcomes*, *17*(3), e010144.

18 Bever, L. (2022). From heart disease to IUDs: How doctors dismiss women's pain. *Washington Post*, Dec. 13.

19 Chen, E., Shofer, F., Dean, A., et al. (2008). Gender disparity in analgesic treatment of emergency department patients with acute abdominal pain. *Academic Emergency Medicine*, *15*(5), 414–418.

20 Otis, J., Keane, T., & Kerns, R. (2003). An examination of the relationship between chronic pain and post-traumatic stress disorder. *Journal of Rehabilitation Research and Development*, *40*(5), 397–406.

21 Egloff, N., Hirschi, A., & von Känel, R. (2013). Traumatization and chronic pain: A further model of interaction. *Journal of Pain Research*, 765–770.

22 Felitti, V. (2002). The relation between adverse childhood experiences and adult health: Turning gold into lead. *Permanente Journal*, *6*(1), 44–47.

23 Felitti, V., Anda, R., Nordenberg, D., et al. (1998). Relationship of childhood abuse and household dysfunction to many of the leading causes of death in adults: The Adverse Childhood Experiences (ACE) Study. *American Journal of Preventive Medicine, 14*(4), 245–258.

24 Monnat, S., & Chandler, R. (2015). Long-term physical health consequences of adverse childhood experiences. *Sociological Quarterly, 56*(4), 723–752.

25 Bussières, A., Hancock, M., Elklit, A., et al. (2023). Adverse childhood experience is associated with an increased risk of reporting chronic pain in adulthood: A systematic review and meta-analysis. *European Journal of Psychotraumatology, 14*(2), 2284025.

26 Tidmarsh, L., Harrison, R., Ravindran, D., et al. (2022). The influence of adverse childhood experiences in pain management: Mechanisms, processes, and trauma-informed care. *Frontiers in Pain Research, 3*, 923866.

27 Lew, H., Poole, J., Vanderploeg, R., et al. (2007). Program development and defining characteristics of returning military in a VA Polytrauma Network Site. *Journal of Rehabilitation Research & Development, 44*(7).

28 Young, D., Chao, L., Neylan, T., et al. (2018). Association among anterior cingulate cortex volume, psychophysiological response, and PTSD diagnosis in a veteran sample. *Neurobiology of Learning and Memory, 155*, 189–196.

29 Fenster, R., Lebois, L., Ressler, K., et al. (2018). Brain circuit dysfunction in post-traumatic stress disorder: From mouse to man. *Nature Reviews Neuroscience, 19*(9), 535–551.

30 Asmundson, G., Coons, M., Taylor, S., et al. (2002). PTSD and the experience of pain: Research and clinical implications of shared vulnerability and mutual maintenance models. *Canadian Journal of Psychiatry, 47*(10), 930–937.

31 Scioli-Salter, E., Forman, D., Otis, J., et al. (2015). The shared neuroanatomy and neurobiology of comorbid chronic pain and PTSD: Therapeutic implications. *Clinical Journal of Pain, 31*(4), 363–374.

32 Bosco, M., Gallinati, J., & Clark, M. (2013). Conceptualizing and treating comorbid chronic pain and PTSD. *Pain Research and Treatment, 2013*(1), 174728.

9. THE BODY'S PHARMACY

1 De la Fuente-Fernández, R., Ruth, T., Sossi, V., et al. (2001). Expectation and dopamine release: Mechanism of the placebo effect in Parkinson's disease. *Science, 293*(5532), 1164–1166.

2. Benedetti, F., Frisaldi, E., Carlino, E., et al. (2016). Teaching neurons to respond to placebos. *Journal of Physiology, 594*(19), 5647–5660.

3. Lidstone, S., Schulzer, M., Dinelle, K., et al. (2010). Effects of expectation on placebo-induced dopamine release in Parkinson's disease. *Archives of General Psychiatry, 67*(8), 857–865.

4. Benedetti, F., Pollo, A., Lopiano, L., et al. (2003). Conscious expectation and unconscious conditioning in analgesic, motor, and hormonal placebo/nocebo responses. *Journal of Neuroscience, 23*(10), 4315–4323.

5. Benedetti, F. (2006). Placebo analgesia. *Neurological Sciences, 27*, s100–s102.

6. Colloca, L., & Benedetti, F. (2006). How prior experience shapes placebo analgesia. *Pain, 124*(1–2), 126–133.

7. Rossettini, G., Camerone, E., Carlino, E., et al. (2020). Context matters: The psychoneurobiological determinants of placebo, nocebo and context-related effects in physiotherapy. *Archives of Physiotherapy, 10*, 1–12.

8. Colloca, L., Klinger, R., Flor, H., et al. (2013). Placebo analgesia: Psychological and neurobiological mechanisms. *Pain, 154*(4), 511–514.

9. Benedetti, F., Carlino, E., & Pollo, A. (2011). How placebos change the patient's brain. *Neuropsychopharmacology, 36*(1), 339–354.

10. Rossettini, G., Campaci, F., Bialosky, J., et al. (2023). The biology of placebo and nocebo effects on experimental and chronic pain: State of the art. *Journal of Clinical Medicine, 12*(12), 4113.

11. Benedetti, F., Shaibani, A., Arduino, C., et al. (2023). Open-label nondeceptive placebo analgesia is blocked by the opioid antagonist naloxone. *Pain, 164*(5):984–990.

12. Moseley, J., O'Malley, K., Petersen, N., et al. (2002). A controlled trial of arthroscopic surgery for osteoarthritis of the knee. *New England Journal of Medicine, 347*(2), 81–88.

13. Louw, A., Diener, I., Fernández-de-Las-Peñas, C., et al. (2017). Sham surgery in orthopedics: A systematic review of the literature. *Pain Medicine, 18*(4), 736–750.

14. Jonas, W., Crawford, C., Colloca, L., et al. (2015). To what extent are surgery and invasive procedures effective beyond a placebo response? A systematic review with meta-analysis of randomised, sham controlled trials. *BMJ Open, 5*(12), e009655.

15. Reeves, R., Ladner, M., Hart, R., et al. (2007). Nocebo effects with antidepressant clinical drug trial placebos. *General Hospital Psychiatry, 29*(3), 275–277.

16 Benedetti, F., Amanzio, M., Vighetti, S., et al. (2006). The biochemical and neuroendocrine bases of the hyperalgesic nocebo effect. *Journal of Neuroscience, 26*(46), 12014–12022.

17 Schedlowski, M., Enck, P., Rief, W., et al. (2015). Neuro-bio-behavioral mechanisms of placebo and nocebo responses: Implications for clinical trials and clinical practice. *Pharmacological Reviews, 67*(3), 697–730.

18 Thomaidou, M., Peerdeman, K., Koppeschaar, M., et al. (2021). How negative experience influences the brain: A comprehensive review of the neurobiological underpinnings of nocebo hyperalgesia. *Frontiers in Neuroscience, 15*, 652552.

19 Benedetti, F., Lanotte, M., Lopiano, L., et al. (2007). When words are painful: Unraveling the mechanisms of the nocebo effect. *Neuroscience, 147*(2), 260–271.

20 Keltner, J., Furst, A., Fan, C., et al. (2006). Isolating the modulatory effect of expectation on pain transmission: A functional magnetic resonance imaging study. *Journal of Neuroscience, 26*(16), 4437–4443.

21 Varelmann, D., Pancaro, C., Cappiello, E., et al. (2010). Nocebo-induced hyperalgesia during local anesthetic injection. *Anesthesia & Analgesia, 110*(3), 868–870.

22 Linskens, F., van der Scheer, E., Stortenbeker, I., et al. (2023). Negative language use of the physiotherapist in low back pain education impacts anxiety and illness beliefs: A randomised controlled trial in healthy respondents. *Patient Education and Counseling, 110*, 107649.

23 Wells, R., & Kaptchuk, T. (2012). To tell the truth, the whole truth, may do patients harm: The problem of the nocebo effect for informed consent. *American Journal of Bioethics, 12*(3), 22–29.

24 Kaptchuk, T., Friedlander, E., Kelley, et al. (2010). Placebos without deception: A randomized controlled trial in irritable bowel syndrome. *PloS one, 5*(12), e15591.

25 Bernstein, M., Fuchs, N., Rosenfield, M., et al. (2021). Treating pain with open-label placebos: A qualitative study with post-surgical pain patients. *Journal of Pain, 22*(11), 1518–1529.

10. WELCOME TO THE REVOLUTION

1 Elbers, S., Wittink, H., Konings, S., et al. (2022). Longitudinal outcome evaluations of Interdisciplinary Multimodal Pain Treatment programmes for patients with chronic primary musculoskeletal pain: A systematic review and meta-analysis. *European Journal of Pain, 26*(2), 310–335.

2   U.S. Department of Health and Human Services. (2020). *Pain Management Best Practices Inter-Agency Task Force Report*. From https://www.hhs.gov/sites/default/files/pmtf-final-report-2019-05-23.pdf.

3   Dowell D., Ragan K., Jones C., et al. (2022). CDC Clinical Practice Guideline for Prescribing Opioids for Pain—United States. *MMWR Recommendations & Reports*, 71(No. RR-3), 1–95.

4   Bujak, B., Regan, E., Beattie, P., et al. (2019). The effectiveness of interdisciplinary intensive outpatient programs in a population with diverse chronic pain conditions: A systematic review and meta-analysis. *Pain Management*, 9(4), 417–429.

5   Gatchel, R., McGeary, D., McGeary, C., et al. (2014). Interdisciplinary chronic pain management: Past, present, and future. *American Psychologist*, 69(2), 119.

6   Randall, E., Smith, K., Conroy, C., et al. (2018). Back to living: Long-term functional status of pediatric patients who completed intensive interdisciplinary pain treatment. *Clinical Journal of Pain*, 34(10), 890–899.

7   Cheatle, M. (2016). Biopsychosocial approach to assessing and managing patients with chronic pain. *Medical Clinics*, 100(1), 43–53.

8   Gatchel, R., & Okifuji, A. (2006). Evidence-based scientific data documenting the treatment and cost-effectiveness of comprehensive pain programs for chronic nonmalignant pain. *Journal of Pain*, 7(11), 779–793.

9   Oslund, S., Robinson, R., Clark, T., et al. (2009). Long-term effectiveness of a comprehensive pain management program: Strengthening the case for interdisciplinary care. *Baylor University Medical Center Proceedings*, 22(3), 211–214.

10 Stanos, S. (2012). Focused review of interdisciplinary pain rehabilitation programs for chronic pain management. *Current Pain and Headache Reports*, 16, 147–152.

11 Mercer Lindsay, N., Chen, C., Gilam, G., et al. (2021). Brain circuits for ain and its treatment. *Science Translational Medicine*, 13(619), eabj7360.

12 Moseley, G., & Flor, H. (2012). Targeting cortical representations in the treatment of chronic pain: A review. *Neurorehabilitation and Neural Repair*, 26(6), 646–652.

11. THE BIOLOGY OF BALANCE

1   Luque-Suarez, A., Martinez-Calderon, J., & Falla, D. (2019). Role of kinesiophobia on pain, disability and quality of life in people suffering from chronic

musculoskeletal pain: A systematic review. *British Journal of Sports Medicine*, *53*(9), 554–559.

2. Mahdavi, S., Riahi, R., Vahdatpour, B., et al. (2021). Association between sedentary behavior and low back pain: A systematic review and meta-analysis. *Health Promotion Perspectives*, *11*(4), 393.

3. Rice, D., Nijs, J., Kosek, E., et al. (2019). Exercise-induced hypoalgesia in pain-free and chronic pain populations: State of the art and future directions. *Journal of Pain*, *20*(11), 1249–1266.

4. Duran, A., Friel, C., Serafini, M., et al. (2023). Breaking up prolonged sitting to improve cardiometabolic risk: Dose–response analysis of a randomized crossover trial. *Medicine & Science in Sports & Exercise*, *55*(5), 847–855.

5. O'Grady, H., Hasan, H., Rochwerg, B., et al. (2024). Leg cycle ergometry in critically ill patients—an updated systematic review and meta-analysis. *NEJM Evidence*, *3*(12), EVIDoa2400194.

6. Haack, M., Simpson, N., Sethna, N., et al. (2020). Sleep deficiency and chronic pain: Potential underlying mechanisms and clinical implications. *Neuropsychopharmacology*, *45*(1), 205–216.

7. Davidson, J., Dickson, C., & Han, H. (2019). Cognitive behavioural treatment for insomnia in primary care: A systematic review of sleep outcomes. *British Journal of General Practice*, *69*(686), e657–e664.

8. Selvanathan, J., Pham, C., Nagappa, M., et al. (2021). Cognitive behavioral therapy for insomnia in patients with chronic pain–a systematic review and meta-analysis of randomized controlled trials. *Sleep Medicine Reviews*, *60*, 101460.

9. Tang, Nicole K. (2021) Is cognitive-behaviour therapy for insomnia (CBT-I) the new best pain killer? *Sleep Medicine Reviews*, *60*, 101536.

10. Robblee, M., Kim, C., Abate, et al. (2016). Saturated fatty acids engage an IRE1α-dependent pathway to activate the NLRP3 inflammasome in myeloid cells. *Cell Reports*, *14*(11), 2611–2623.

11. Elma, Ö., Brain, K., & Dong, H. (2022). The importance of nutrition as a lifestyle factor in chronic pain management: A narrative review. *Journal of Clinical Medicine*, *11*(19), 5950.

12. Field, R., Pourkazemi, F., Turton, J., et al. (2021). Dietary interventions are beneficial for patients with chronic pain: A systematic review with meta-analysis. *Pain Medicine*, *22*(3), 694–714.

13. Edwards, S., Martin, S., Rainey, T., et al. (2023). Influence of acute fasting on pain tolerance in healthy subjects: A randomised crossover study. *Frontiers in Pain Research*, *4*, 1153107.

14 Offit, P. (2013). The vitamin myth: Why we think we need supplements. *The Atlantic*, July 19.

15 Chou, R., Turner, J., Devine, et al. (2015). The effectiveness and risks of long-term opioid therapy for chronic pain: A systematic review for a National Institutes of Health Pathways to Prevention Workshop. *Annals of Internal Medicine, 162*(4), 276–286.

16 Cashin A., Wand, B., O'Connell, N., et al. (2023). Pharmacological treatments for low back pain in adults: An overview of Cochrane Reviews. *Cochrane Database of Systematic Reviews, 4*.

17 Gan T. (2017). Poorly controlled postoperative pain: Prevalence, consequences, and prevention. *Journal of Pain Research*, 2287–2298.

12. THE EMOTIONAL HEALTH PROTOCOL

1 Toledo, T., Vore, C., Huber, F., et al. (2024). The effect of emotion regulation on the emotional modulation of pain and nociceptive flexion reflex. *Pain, 165*(6), 1266–1277.

2 Greenwald, J., & Shafritz, K. (2018). An integrative neuroscience framework for the treatment of chronic pain: From cellular alterations to behavior. *Frontiers in Integrative Neuroscience, 12*, 18.

3 Shi, J., Liu, Z., Zhou, X., et al. (2023). Effects of breathing exercises on low back pain in clinical: A systematic review and meta-analysis. *Complementary Therapies in Medicine, 79*, 102993.

4 Kondo, K., Noonan, K., Freeman, M., et al. (2019). Efficacy of biofeedback for medical conditions: An evidence map. *Journal of General Internal Medicine, 34*, 2883–2893.

5 Zeidan, F., & Vago, D. R. (2016). Mindfulness meditation–based pain relief: A mechanistic account. *Annals of the New York Academy of Sciences, 1373*(1), 114–127.

6 Zeidan, F., Martucci, K., Kraft, R., et al. (2011). Brain mechanisms supporting the modulation of pain by mindfulness meditation. *Journal of Neuroscience, 31*(14), 5540–5548.

7 Cross, M., Acevedo, A., Leger, K., et al. (2023). How and why could smiling influence physical health? A conceptual review. *Health Psychology Review, 17*(2), 321–343.

8 Manninen, S., Tuominen, L., Dunbar, R., et al. (2017). Social laughter triggers endogenous opioid release in humans. *Journal of Neuroscience, 37*(25), 6125–6131.

9   Müller, R., Terrill, A., Jensen, M., et al. (2015). Happiness, pain intensity, pain interference, and distress in individuals with physical disabilities. *American Journal of Physical Medicine & Rehabilitation*, *94*(12), 1041–1051.

10 Leknes, S., & Tracey, I. (2008). A common neurobiology for pain and pleasure. *Nature reviews neuroscience*, *9*(4), 314–320.

11 Pennebaker, J. (1997). Writing about emotional experiences as a therapeutic process. *Psychological Science*, *8*(3), 162–166.

12 Bao, S., Qiao, M., Lu, Y., et al. (2022). Neuroimaging mechanism of cognitive behavioral therapy in pain management. *Pain Research and Management*, *1*, 6266619.

13 Cunningham, N., Kashikar-Zuck, S., & Coghill, R. (2019). Brain mechanisms impacted by psychological therapies for pain: Identifying targets for optimization of treatment effects. *Pain reports*, *4*(4), e767.

14 Majeed, M., & Sudak, D. (2017). Cognitive behavioral therapy for chronic pain—One therapeutic approach for the opioid epidemic. *Journal of Psychiatric Practice*, *23*(6), 409–414.

15 Tick, H., Nielsen, A., Pelletier, K., et al. (2018). Evidence-based nonpharmacologic strategies for comprehensive pain care: The consortium pain task force white paper. *Explore*, *14*(3), 177–211.

16 O'Sullivan, P., Caneiro, J., O'Keeffe, M., et al. (2018). Cognitive Functional Therapy: An integrated behavioral approach for the targeted management of disabling low back pain. *Physical Therapy*, *98*(5), 408–423.

17 Hollon, S., DeRubeis, R., Andrews, P., et al. (2021). Cognitive therapy in the treatment and prevention of depression: A fifty-year retrospective with an evolutionary coda. *Cognitive Therapy and Research*, *45*(3), 402–417.

13. THE POWER OF CHANGING YOUR MIND

1   Stewart, M., & Loftus, S. (2018). Sticks and stones: The impact of language in musculoskeletal rehabilitation. *Journal of Orthopaedic & Sports Physical Therapy*, *48*(7), 519–522.

2   Bedell, S., Graboys, T., Bedell, E., et al. (2004). Words that harm, words that heal. *Archives of Internal Medicine*, *164*(13), 1365–1368.

3   Kent, P., Haines, T., O'Sullivan, P., et al. (2023). Cognitive Functional Therapy with or without movement sensor biofeedback versus usual care for chronic, disabling low back pain (RESTORE): A randomised, controlled, three-arm, parallel group, phase 3, clinical trial. *Lancet*, *401*(10391), 1866–1877.

4   Murphy, J., Cordova, M., & Dedert, E. (2022). Cognitive behavioral therapy for chronic pain in veterans: Evidence for clinical effectiveness in a model program. *Psychological Services*, *19*(1), 95.

5   Rozanski, A., Bavishi, C., Kubzansky, L., et al. (2019). Association of optimism with cardiovascular events and all-cause mortality: A systematic review and meta-analysis. *JAMA Network Open*, *2*(9), e1912200.

6   Forte, A., Guliyeva, G., McLeod, H., et al. (2022). The impact of optimism on cancer-related and postsurgical cancer pain: A systematic review. *Journal of Pain and Symptom Management*, *63*(2), e203–e211.

7   Hanssen, M., Peters, M., Vlaeyen, J., et al. (2013). Optimism lowers pain: Evidence of the causal status and underlying mechanisms. *Pain*, *154*(1), 53–58.

8   Kılıç, A., Hudson, J., McCracken, L., et al. (2021). A systematic review of the effectiveness of self-compassion-related interventions for individuals with chronic physical health conditions. *Behavior Therapy*, *52*(3), 607–625.

9   Edwards, K., Pielech, M., Hickman, et al. (2019). The relation of self-compassion to functioning among adults with chronic pain. *European Journal of Pain*, *23*(8), 1538–1547.

10 Williams, N. (2009). Words that harm: Words that heal. *International Musculoskeletal Medicine*, *31*(3), 99–100.

11 Moffett, J., Green, A., & Jackson, D. (2013). Words that help, words that harm. In *Topical Issues in Pain 5* (pp. 105). CNS Press.

12 Webster, B., Bauer, A., Choi, Y., et al. (2013). Iatrogenic consequences of early magnetic resonance imaging in acute, work-related, disabling low back pain. *Spine*, *38*(22), 1939–1946.

13 Hall, A., Aubrey-Bassler, K., Thorne, B., et al. (2021). Do not routinely offer imaging for uncomplicated low back pain. *BMJ*, *372*.

14 Ogino, Y., Nemoto, H., Inui, K., et al. (2007). Inner experience of pain: Imagination of pain while viewing images showing painful events forms subjective pain representation in human brain. *Cerebral Cortex*, *17*(5), 1139–1146.

15 Moseley, G., Zalucki, N., Birklein, F., et al. (2008). Thinking about movement hurts: The effect of motor imagery on pain and swelling in people with chronic arm pain. *Arthritis Care & Research*, *59*(5), 623–631.

16 Trakhtenberg, E. (2008). The effects of guided imagery on the immune system: A critical review. *International Journal of Neuroscience*, *118*(6), 839–855.

17 Kaplun, A., Alperovitch-Najenson, D., & Kalichman, L. (2021). Effect of guided imagery on pain and health-related quality of life in musculoskeletal medicine: A comprehensive narrative review. *Current Pain and Headache Reports*, *25*(12), 76.

18 Silvestrini, N., & Corradi-Dell'Acqua, C. (2023). Distraction and cognitive control independently impact parietal and prefrontal response to pain. *Social Cognitive and Affective Neuroscience, 18*(1), nsad018.

19 Subnis, U., Starkweather, A., & Menzies, V. (2016). A current review of distraction-based interventions for chronic pain management. *European Journal of Integrative Medicine, 8*(5), 715–722.

20 Pergolizzi Jr, J., Coluzzi, F., Colucci, R., et al. (2020). Statins and muscle pain. *Expert Review of Clinical Pharmacology, 13*(3), 299–310.

21 Fusco, N., Bernard, F., Roelants, F., et al. (2020). Hypnosis and communication reduce pain and anxiety in peripheral intravenous cannulation: Effect of language and confusion on pain during peripheral intravenous catheterization (KTHYPE), a multicentre randomised trial. *British Journal of Anaesthesia, 124*(3), 292–298.

22 Elkins, G., Jensen, M., & Patterson, D. (2007). Hypnotherapy for the management of chronic pain. *International Journal of Clinical and Experimental Hypnosis, 55*(3), 275–287.

23 Licciardone, J., Tran, Y., Ngo, K., et al. (2024). Physician empathy and chronic pain outcomes. *JAMA Network Open, 7*(4), e246026–e246026.

14. SOCIAL MEDICINE

1 Pilkington, G., Johnson, M., & Thompson, K. (2025). Social prescribing for adults with chronic pain in the UK: A rapid review. *British Journal of Pain*, 20494637241312064.

2 Ashton-James, C., Anderson, S., Mackey, S., et al. (2022). Beyond pain, distress, and disability: The importance of social outcomes in pain management research and practice. *Pain, 163*(3), e426–e431.

3 Carey, B., Dell, C., Stempien, J., et al. (2022). Outcomes of a controlled trial with visiting therapy dog teams on pain in adults in an emergency department. *PLoS One, 17*(3), e0262599.

4 Zhang, Y., Yan, F., Li, S., et al. (2021). Effectiveness of animal-assisted therapy on pain in children: A systematic review and meta-analysis. *International Journal of Nursing Sciences, 8*(1), 30–37.

5 Havey, J., Vlasses, F., Vlasses, P., et al. (2014). The effect of animal-assisted therapy on pain medication use after joint replacement. *Anthrozoös, 27*(3), 361–369.

6 Wurm, M., Edlund, S., Tillfors, M., et al. (2016). Characteristics and consequences of the co-occurrence between social anxiety and pain-related fear

in chronic pain patients receiving multimodal pain rehabilitation treatment. *Scandinavian Journal of Pain*, *12*(1), 45–52.

7   Reczek, R., Stacey, L., & Thomeer, M. (2023). Parent–adult child estrangement in the United States by gender, race/ethnicity, and sexuality. *Journal of Marriage and Family*, *85*(2), 494–517.

8   Teoli, D., Dua, A., & An, J. (2024). Transcutaneous electrical nerve stimulation. In *StatPearls [Internet]*. StatPearls Publishing.

9   Vance C., Dailey D., Rakel B., et al. (2014). Using TENS for pain control: The state of the evidence. *Pain Management*, *4*(3):197–209.

10 Simoncini, E., Stiaccini, G., Morelli, E., et al. (2023). The Effectiveness of the Buzzy Device in Reducing Pain in Children Undergoing Venipuncture: A Single-Center Experience. *Pediatric Emergency Care*, *39*(10), 760–765.

11 Şahan, S., & Yildiz, A. (2022). The effect of shotblocker application on intramuscular injection pain in adults: A meta-analysis. *Clinical Nursing Research*, *31*(5), 820–825.

12 Dusek, J., Griffin, K., Finch, M., et al. (2018). Cost savings from reducing pain through the delivery of integrative medicine program to hospitalized patients. *Journal of Alternative and Complementary Medicine*, *24*(6), 557–563.

13 Bosco, M., Gallinati, J., & Clark, M. (2013). Conceptualizing and treating comorbid chronic pain and PTSD. *Pain Research and Treatment*, *2013*(1), 174728.

14 Lumley, M., Yamin, J., Pester, B., et al. (2022). Trauma matters: Psychological interventions for comorbid psychosocial trauma and chronic pain. *Pain*, *163*(4), 599–603.

15 Norman, S., Wilkins, K., Tapert, S., et al. (2010). A pilot study of Seeking Safety therapy with OEF/OIF veterans. *Journal of Psychoactive Drugs*, *42*(1), 83–87.

16 Gainer, D., Alam, S., Alam, H., et al. (2020). A flash of hope: Eye movement desensitization and reprocessing (EMDR) therapy. *Innovations in Clinical Neuroscience*, *17*(7–9), 12.

17 Watkins, L., Sprang, K., & Rothbaum, B. (2018). Treating PTSD: A review of evidence-based psychotherapy interventions. *Frontiers in Behavioral Neuroscience*, *12*, 258.

18 Kuhfuß, M., Maldei, T., Hetmanek, A., et al. (2021). Somatic experiencing–effectiveness and key factors of a body-oriented trauma therapy: A scoping literature review. *European Journal of Psychotraumatology*, *12*(1), 1929023.

19 Van der Kolk, B. (1994). The body keeps the score: Memory and the evolving

psychobiology of posttraumatic stress. *Harvard Review of Psychiatry, 1*(5), 253–265.

20 Schnurr, P., Hamblen, J., Wolf, J., et al. (2024). The management of posttraumatic stress disorder and acute stress disorder: Synopsis of the 2023 US Department of Veterans Affairs and US Department of Defense clinical practice guideline. *Annals of Internal Medicine, 177*(3), 363–374.

21 Hertzberg, M., Feldman, M., Beckham, J., et al. (2000). Lack of efficacy for fluoxetine in PTSD: A placebo-controlled trial in combat veterans. *Annals of Clinical Psychiatry, 12*, 101–105.

22 Perry, B., & Winfrey, O. (2021). *What Happened to You? Conversations on Trauma, Resilience, and Healing.* Flatiron Books.

About the Author

Dr. Rachel Zoffness MS, PhD, is a leading global pain expert, pain psychologist, medical consultant, speaker, author, disruptor, and thought leader in pain medicine. She's faculty at the University of California San Francisco School of Medicine, lectures at Stanford, and is a winner of the prestigious Mayday Fellowship.